RIFLE ACCURACY FACTS

A distinguished scientist's lifelong pursuit of the secrets of "Extreme rifle accuracy"

by
Harold R. Vaughn

Precision
SHOOTING
Manchester, Connecticut USA

Published by:

Precision Shooting, Inc.
222 McKee Street
Manchester, Connecticut 06040

Phone: (860) 645-8776
Fax: (860) 643-8215

MCMXCVIII

Second Edition, January 2000

ACKNOWLEDGMENTS

The following is a partial list of people that have contributed technical expertise and editing help to the author.

Ed Adams (Civil Engineer) - Bench rest shooting, Tunnel Range.

Roy Baty (PhD, Aero) - Mathematics, jet flow, Tunnel Range

W. T. Atkinson (Custom Barrel Maker) - Barrel work.

Harold Bennett (MS, EE) - Electronics, bench rest shooting.

James K. Cole (PhD, Aero) - Shadowgraph photography.

Robert Croll (MS, EE) - Shadowgraph technology.

Frank A. Hemsted (Bullet Die Maker - deceased) - Die and reamer machine work.

Jack E. Jackson (MD, PhD Chemistry) - Benchrest accuracy problems, chemistry, shadowgraph work, primary editor.

Walter Jankowsky (Cook Bullets, deceased) - Bullet making, rail guns.

A.A. Leiber (MS, EE, deceased) - Accuracy problems, editing.

George Reis (Physicist) - Internal ballistics, instrumentation.

Frank Tirrell (Gunsmith) - Barrel making, rail gun technology, benchrest shooting.

Mark Vaughn (PhD ME) - Structural design, thread design, Tunnel Range.

Leslie Vaughn (MS Chemistry) - Chemistry.

Zia Rifle Club members that helped in the construction of the Tunnel Range (Stan Barnhart, John Winder, Allan Rittgers, Richard Henderson, Dick Vivian, and others), and Bill White.

Dedication

The book is dedicated to
my wife Mary
who supported all this work
and edited the rough draft.

ABOUT THE AUTHOR

Harold Roy Vaughn was born in 1924 at the family farm a few miles south of Amarillo, Texas. After graduating from high school, he entered Amarillo Junior College in September, 1941, to study engineering. He volunteered for the Army Air Corps Reserve (the beginning of the US Air Force) in June, 1942, and reported for duty in February, 1943. He flew 100 combat missions in P-47's and P-51's from bases in New Guinea, Morotai, the Philippines, China, and Okinawa and was awarded the Air Medal with four Oak Leaf clusters and seven battle stars during his tour of duty. Colonel Charles Lindbergh flew several missions with Harold's squadron as a civilian technical consultant to demonstrate how to obtain more aircraft range with optimum throttle and propeller speed settings. Harold returned to civilian life in January, 1946, and to Amarillo Junior College to finish the last semester of his sophomore year in engineering. During the summer of 1946, he entered the University of Colorado and received a BS in aerodynamics in 1948 and a MS in aerodynamics in 1949. He worked at the NACA (now NASA) Ames Research Laboratory, Moffett Field, California, from September 1949 to September 1951, where he conducted research on the aerodynamics of swept wings. In September 1951 he joined Sandia National Laboratories, Albuquerque, New Mexico as a staff member in the Aerodynamics Department. He was promoted to Supervisor, Aeroballistics Division in July 1959, a position he held until his retirement from Sandia National Laboratories (SNL) in 1986. This division provided the flight dynamics and the aerodynamic research and development and design for nuclear weapons. As supervisor of this division, he provided technical direction to a large staff of scientists.

Harold is considered the "grandfather" of the aeroballistics/flight mechanics technology base for nuclear ordnance at SNL. In the early 1950's he recognized the ballistics problem of roll-pitch resonance of tactical bombs. He mathematically modeled this motion and then recommended a fix of fin tabs, canted fins, or spin rockets to spin through the bomb pitch frequency, thereby avoiding divergent pitch/yaw motion. These solutions have been used on all nuclear bombs and sounding rocket systems at SNL. He was responsible for the aerodynamic design of a rocket boosted Mach 5 test vehicle to test baro-fuzing probes in 1957. He pioneered the use of computers in the field to calculate launcher settings to minimize dispersion for the several hundred unguided instrumentation rockets launched at Kauai and Johnston Island during the 1958 and 1962 high altitude nuclear tests. He was responsible for the aerodynamic design of the 14,000 pound Strypi rocket system which was developed in the fall of 1962 to boost a 560 kilogram nuclear warhead to an altitude of 150 kilometers at Johnston Island for the Checkmate event of the Dominic high altitude test series. He developed and published theories for analyzing reentry vehicle motion and dispersion, including the effects of roll resonance, heat shield thermal distortion, aerodynamic or inertial asymmetries, spinup, and exoatmospheric nuclear attack. One of these publications is extensively quoted in F. J. Regan's 1984 book Re-Entry Vehicle Dynamics. He developed and published a theory for the ballistic match (same impact point for identical launch conditions) of nuclear and high explosive warhead artillery shells. This theory identified the required matching inertial parameters which enabled the Los Alamos National Laboratory and the Lawrence Livermore National Laboratories to design the nuclear warheads. He developed and published a comprehensive theory for calculating forces and moments on spinning shells. In the early 1980's he published the numerical solution of the Navier-Stokes equations to predict the fluid motion inside a spinning nutating cylinder using the Cray super-computer. This first theoretical solution explained the flight instabilities of spin-stabilized, liquid-filled artillery shells. He also initiated many other programs such as (1) pioneering the use of computers to obtain complete trajectory calculations for bombs, shells, rockets, reentry vehicles, etc. (2) working with the SNL Field Test organization to develop a miniaturized

3-axis system for use on-board test vehicles to measure angular motion which is telemetered to a ground station and (3) conceiving of and developing the SNL Flight Simulation Laboratory.

Harold received the 1974 American Institute of Aeronautics and Astronautics Mechanics and Control of Flight Award. The award was "for his fundamental contributions to the understanding of the flight mechanics of reentry vehicles, rockets, bombs and shells, together with his innovations in their aerodynamic design for stability and minimum dispersion in transonic flight." He received the Outstanding Civilian Service Award from the U. S. Department of the Army in 1976 for solving a serious ballistics problem with the M422 shell. Movement of parts inside the shell caused a large undamped nutational motion that increased the drag, thereby markedly decreasing the range. He received the Department of Energy "Award of Excellence" from Major General W. W. Hoover in 1982 for significant contributions to the nuclear weapons program for "Ballistic Similitude" of artillery shells.

Harold has many hobbies — big game hunting, oil painting, photography, electronics, skiing, fly fishing, gardening, ultralight aircraft, and precision shooting. His advancing years have made some of these hobbies fond memories but he still pursues the less physically demanding ones. Behind his desk in his spacious study hangs a majestic elk, originally number 13 in the book. A grand slam on sheep adorns the fireplace wall. Numerous other big game trophies decorate the study. A small, well equipped photographic darkroom opens off of his study. A short hallway leads to a shop at the back of his garage that contains a Clausing lathe and vertical mill plus numerous other pieces of equipment. This well equipped shop is constantly used for projects related to precision shooting.

Jack E. Jackson

TABLE OF CONTENTS

Chapter **Page**

CHAPTER 1
INTRODUCTION

Some forty five years ago, when I started big game hunting, I became dissatisfied with the accuracy of commercial rifles. You just don't want to miss after spending days and sometimes weeks looking for a big trophy, and then finally getting one shot at three hundred yards or more. Most sporting rifles are not accurate enough for these long range shots. The commercial rifles that I tested would shoot 5 shot groups ranging from 4 inches to 12 inches at 300 yards, and that just isn't good enough for a serious trophy hunter.

Now, a lot of you will say that your rifle is capable of shooting more accurately than you are capable of shooting. Now I'll buy that, if you happen to be one of those people that just can't shoot because of flinching, or not being able to see well, or for some other reason. However, I can't agree with this for the majority of shooters, because I have fired thousands of rounds through accurate sporters on machine rests where the only skill involved is putting the cross hairs of a 20 power scope on the center of the target. Invariably, I get about the same accuracy when I, and other folks, shoot the same gun from the shoulder at prone position or from a bench rest. Bench rest shooters have been consistently shooting better than 0.3 inch 10 shot groups at 100 yards for years with specially made heavy rifles and carefully assembled ammunition, while it is rare for a sporter to shoot better than 1.5 inch 5 shot groups at 100 yards. This should be ample proof that most people can shoot a lot better than their guns are capable of shooting. By the early 1960's I had light weight sporters that would reliably shoot 2.5 inch groups at 300 yards, which is adequate for any big game hunting. This was done by replacing the barrel with a custom barrel chambered with my homemade reamers and by replacing the stock with a carefully inletted stock.

So, from a purely practical point of view, the hunting rifle accuracy problem was solved as far as I was concerned. However, I couldn't quit while I was ahead of the game, because of my natural curiosity as an engineer and professional ballistician. It is the same incentive that drives the bench rest shooter to put them all through the same hole in a target, except that I was much more interested in why bullets didn't go through the same hole than doing it in competition. The bench rest people and custom barrel and action people made improvements almost exclusively by trial and error—when something works don't change it. While this is a fairly successful approach in the end, it doesn't really answer the questions in a factual manner nor suggest which factors are more or less important. With very rare exceptions, you can't find anything in the literature where people have really measured anything and pinned something down so that you can be positive about it. It is usually based on someone's guess, which may or may not be right. Unfortunately, most custom gunsmiths and shooters aren't equipped to do anything but group testing and are unaware of the difficulty of statistically isolating small errors. And why haven't modern experimental and computer techniques been applied to rifle accuracy problems when they have been available for years? First off the military are not particularly interested in gilt edge accuracy in rifles. They are interested in effectiveness, which means reliability, rate of fire, cost, and weight (logistics). They are very interested in accuracy in the big bore stuff (cannons), but the problems are somewhat different. The rifle manufacturers are undoubtedly interested, but let's face it, many shooters probably buy a rifle for reasons other than accuracy, and this kind of research would be very expensive when done in a large research laboratory. Consequently, the only way that this thing can get done is for someone to do it on an amateur basis (no pay), who has an extensive background and experience in internal and external ballistics, electronics, mechanical design, machine work, shop equipment, shooting, and a lot of free time. Well, I retired after 37 years of solving all types of projectile problems and decided to take a shot at it.

The general approach in this book is to report experimental measurements that are conclusive wherever possible, and where it is not possible, theory and computer solutions are used. Guesses as to the cause of some phenomena that defy obtaining conclusive evidence will occasionally be given, but they will be clearly stated as guesses. While the author is certain that most of the major causes of inaccuracy have been isolated, evaluated and minimized by redesign in this work, it is also clear that every problem has not been solved. An attempt has been made to write this book so that it can be understood by those with little or no formal technical training. Don't be dismayed by the few equations that have been included for the benefit of my scientific colleagues that may want to know exactly what was done. All theory is explained in simple physical language, so you can skip the equations and still understand what has been done.

The reader should note that I have used the words accuracy, precision, and group size in this book. Strictly speaking they mean different things. Accuracy describes the ability of a rifle to hit a given spot on a target, while precision and group size means the ability to shoot a small group any place on a target. Precision is a term used by bench rest shooters to describe a small group. I have used these terms somewhat indiscriminately because I feel that you can't have accuracy without having a small group size.

Before getting into the nitty-gritty of rifle accuracy, we need to have some rough idea of the attainable accuracy of commercial sporter rifles in reasonable weights. I consider any rifle over eight pounds to be too heavy to be considered a standard sporter. Table 1 shows the data on the accuracy of several rifles that was obtained from the "Rifleman" and other references over a period of years. The data are the results of several 5-shot groups, and the most accurate load is presented for each rifle. The rifles have been divided into four classes-standard sporter, heavy sporter, bench rest and rail gun. The main point is that the typical maximum group size is about 2 inches and the typical average group size is about 1.5 inches for a standard sporter. My experience with this class of rifle, which started in the middle 1940's, indicates that this level of performance is typical. It is also typical of the Remington 721 that I have chosen for the investigation. The table also indicates that heavy rifles are usually more accurate than light ones, which doesn't surprise me, and shouldn't surprise the reader. It also indicates that some of these cartridges, such as the 6mm PPC, are more accurate than others.

During the course of my research we will discover the cause of most of these differences. I wouldn't get too excited about the either good or bad performance of a particular sporter in Table 1, because you have to remember that these rifles were fired by different people using different sights and ammunition under different conditions. The data should be viewed as simply a rough indication of what to expect.

TABLE 1

Summary Of Commercial Rifle Accuracy
(5 shot group size at 100 yards)

Rifle	Cartridge	Weight (pounds)	Date	Group Size (inches) Max.	Avg.
Standard Sporter					
Antonio Zoli, AZ1900	308 Win	7.5	7/90	2.02	1.84
Winchester, M70	270 Wby	7.9	1/89	2.04	1.54
BSA, CF2	30-06	7.9	2/83	2.01	1.51
McMillan Signature	308 Win	7.5	5/88	1.40	.90
Dumoulin Diane	270 Win	7.5	6/88	2.93	2.28
Steyer-Mannlicher	7x64	7.5	7/88	1.52	1.36
Sako, PPC Repeater	6mmPPC	6.5	4/89	1.33	.96
Weatherby, VGX	270 Wby	7.7	2/90	2.36	2.03
Ruger, Mod 77	223 Rem	6.6	12/89	1.73	1.37
			Average	1.93	1.53
Heavy Sporter					
Sako, Six	6mmPPC	8.5	1/88	1.00	.50
Winchester, M70	222 Rem	9.4	4/90	1.52	1.19
Parker Hale, Mod 87	308 Win	11.2	3/88	1.81	1.30

Heavy Varmint Custom Gun (Bench Rest)	
Kelby Action, Hart Barrel by Jim Borden, 6mm PPC, 13.5#	0.20
Rail Gun Unlimited Class	
Remington Action, Shilen Barrel by Vaughn, 6mm BR, 90#	0.18

The reader should note that the data show that while a heavy rifle is likely to shoot better than a light one, weight is not an overriding factor. The thing that makes the most consistent difference is the cartridge that is used. Notice that the 6mm PPC, 6mm BR, 222 Remington, and the 223 Remington generally perform better than the other cartridges. We will find that this is a result of these cartridges having a smaller case diameter, which reduces bolt thrust force, and in addition they use lighter bullets, which results in less recoil force. Both bolt thrust and recoil force cause inaccuracy through barrel vibration. The case diameter of these small cases is 0.378 inches as compared to 0.473 inches for the standard cases, and 0.532 inches for the magnum cases. The smaller bench rest and varmint cases use a faster burning powder, which reduces the effect of muzzle blast on accuracy.

We are going to start with a standard Remington 721 action and stock that has been rebarreled with a Douglas Premium barrel and chambered in the 270 Winchester cartridge. The 270 was chosen, because it is not noted for its accuracy and is a very commonly used cartridge in the medium case capacity and medium recoil range. The Remington 721 was used, because it was available, light, simple, strong, and has a cylindrical receiver, which is best for instrumentation. We start with a custom barrel, because the effects of thermal drift, nonconcentric chamber, and poor throat design can be minimized immediately. Every commercial rifle that the author has checked has had chamber and throat defects, which cause inaccuracy. These causes are discussed in detail in the text, and there is no reason to start with something that you know can cause poor performance. Later, when the research is complete, we will modify an ordinary Remington 721 in 270 Winchester and show that an ordinary rifle will shoot nearly as well as a varmint rifle. You bench rest shooters don't get upset because the problems found in the 270 also occur in bench rest guns, except that these effects are larger in a 270 sporter and are easier to measure.

At the end we will not only have an extremely accurate rifle, but we will understand most of the causes of inaccuracy and how to fix them. Of course, some accuracy problems remain to be solved and I expect to continue working on them. I hope the reader learns as much as I did in doing this work. That would make all the work worthwhile.

The author recommends against trying the experiments shown in this book, because they could be dangerous. While the author has no reason to believe that any of the modifications made to the action or barrel cannot be safely applied in production, not enough experience has been accumulated to be sure that they are safe. All firing tests were made by remotely firing from a machine rest, which is the only safe way to conduct research of this type.

Don't fire an experimental rifle from the shoulder, because it could kill or maim you.

CHAPTER 2 INTERNAL BALLISTICS

The best place to start in trying to isolate and evaluate the various causes of rifle inaccuracies is with the internal ballistics. That is, the ignition of the powder by the primer followed by the generation of chamber pressure by the burning powder, and the travel of the bullet down the barrel. When we get into the nitty-gritty of all these inaccuracy problems and try to either eliminate or minimize their effect, we are going to need to know all about the internal ballistics.

Chamber Pressure Measurement

We have to have a measured chamber pressure to determine all of the interior ballistics quantities that we need to know. There are three available ways to measure chamber pressure, (1) crusher gage, (2) piezoelectric gage, and (3) strain gage. Each of these methods is described.

(1) Crusher Gage - The crusher gage approach involves drilling a hole into the chamber which is threaded for a small cylinder with an inside diameter of about 0.2 inches. A steel piston is dropped into the cylinder, which is followed by the crusher gage and a top threaded cap that restrains the whole thing. The crusher gage is a small copper cylinder which is compressed by the piston being acted on by the chamber pressure.

The chamber pressure is enough to puncture a hole in the wall of the cartridge case. The crusher is calibrated in a static test machine by applying a known compressive force and measuring the amount that the copper cylinder compresses. We won't use this method because it is complicated, requires a lot of precision machine work, destroys the rifle for anything other than pressure testing, gives only the peak chamber pressure where we want a time history of the pressure, and its dynamic accuracy is very doubtful. Much of the pressure data shown in the older reloading books was obtained by this method, and is usually labeled CUP for either crusher or copper units of pressure.

(2) Piezoelectric Gages - This is a newer method that is superior to the crusher gage. However, it requires that a pressure port be drilled into the chamber. Also, they are likely to be sensitive to gun acceleration and temperature, and, most important, they are expensive. The piezoelectric gage contains a small ceramic crystal that generates an electric signal when squeezed by the chamber pressure. This electrical voltage is proportional to the pressure and is recorded on an oscilloscope.

Figure 2-1 - Photograph of the field test setup showing the rifle mounted on a machine rest and the electronic instrumentation.

(3) Strain Gage - The strain gage is the best for our purposes, because it is cheap, has a fast linear response, is nondestructive requiring no machining, gives a time history of the pressure, and can be made insensitive to temperature changes. One problem is that it is difficult to calibrate, but a way has been developed that will be described in detail. A strain gage is a small piece of metal foil that is bonded to the outside of the barrel over the chamber. The internal chamber pressure causes the barrel to expand slightly, which also stretches the strain gage. When the gage is stretched its electrical resistance also changes. This change in resistance can be measured by connecting it in an electrical bridge. The new Oehler Model 43 Ballistics Laboratory uses this principle.

Strain Gage Chamber Pressure Measurement

Figure 2-1 shows the field test setup used in obtaining much of the experimental data presented in this book. The rifle is held in a machine rest that is clamped to the tailgate of the truck. The oscilloscope is located in the right front of the bed, and a portable generator is in front of the truck. Figure 2-2 shows the experimental rifle mounted on a machine rest. The rifle is held by two dovetail slides which allow it to recoil when it is remotely fired by pulling a string. The wires are connected to the strain gages mounted on the chamber section of the barrel and the forward receiver ring. The gages on the forward receiver ring are used to measure moment and will be discussed later (Chapter 4). The location of the strain gages for measuring pressure are shown in Figure 2-3 where it can be seen that the two gages are mounted about 0.4 inches ahead of the forward receiver ring on the cylindrical section of the barrel, which is where the most expansion will occur. If you look

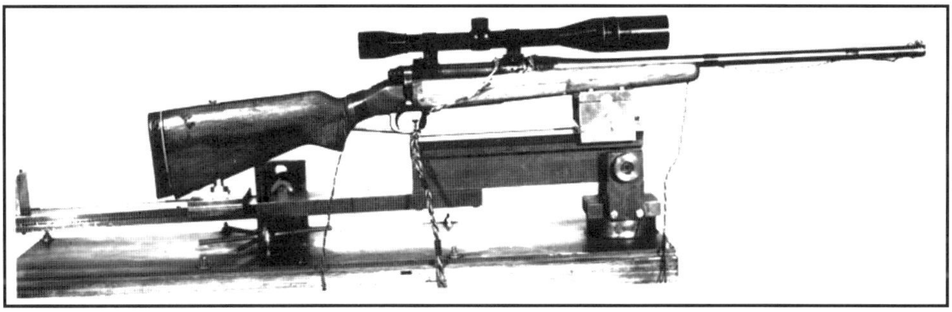

Figure 2-2 - Experimental rifle mounted on the machine rest.

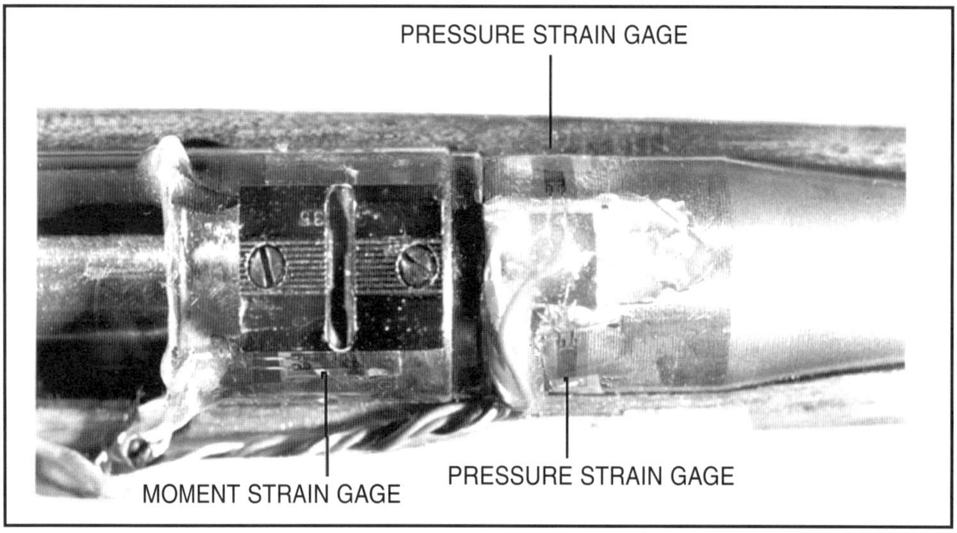

Figure 2-3 - Photograph of the forward receiver ring and barrel showing two strain gages mounted on the barrel chamber for measuring chamber pressure and one of four strain gages on the receiver ring for measuring moment.

carefully, you can also see one of the four moment gages mounted on the receiver ring just below the scope sight mounting block. When the chamber is pressurized it expands circumferentially and longitudinally, however the circumferential expansion or strain is much greater than the longitudinal strain, so we will measure the circumferential strain. A photograph of a strain gage is shown in Figure 2-4. The active part of the gage measures about 1/4 X 1/8 inches. The gage is a very thin metal foil that increases in electrical resistance when it is stretched. This change in resistance can be converted to a voltage change by a strain gage bridge (Figure 2-5). A strain gage bridge is nothing but two resistors and two strain gages, which are also resistors, connected together. Six lantern batteries connected in parallel are connected across the bridge from top to bottom to provide a 6 volt reference voltage.

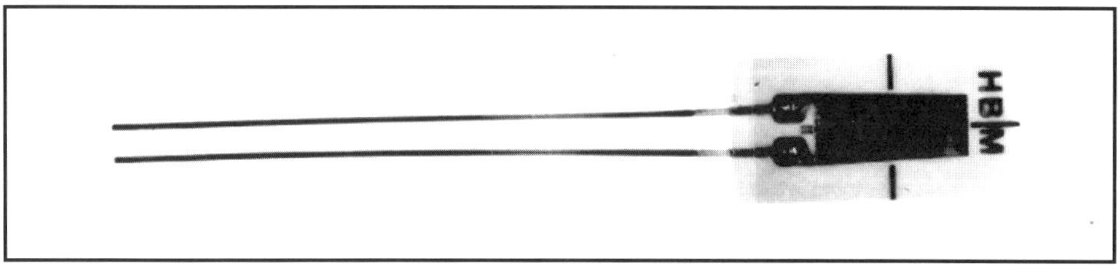

Figure 2-4 - Photograph of a strain gage similar to those used in chamber pressure and receiver ring moment experimental measurements. Actual size is 0.25 by 0.50 inches.

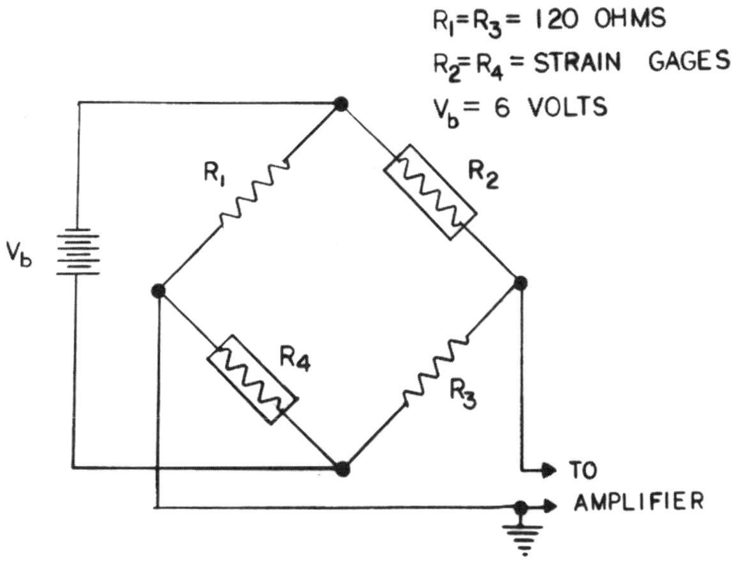

Figure 2-5 - Circuit diagram of strain gage bridge used in chamber pressure measurement.

When the two gages are stretched, the bridge is unbalanced changing the voltage at the two output terminals. Two strain gages are used because this doubles the sensitivity of the measurement and improves the accuracy. The voltage change can be displayed on an oscilloscope (Figure 2-6) and photographed, after it has been amplified. This provides a permanent record of voltage in terms of centimeters (cm) of deflection versus time. The particular oscilloscope (scope for short) is a Tektronix 555 (Figure 2-6), which has the capability of displaying two traces simultaneously. So, if we know the amount of scope trace vertical deflection for a given amount of chamber pressure, we have a direct measurement of chamber pressure as it varies with time. The horizontal deflection with time is done by the internal scope circuits. The usual approach is to use a theoretical calculation of the amount of strain for a given chamber pressure. However this is subject to a large error, and a more accurate experimental calibration will be used.

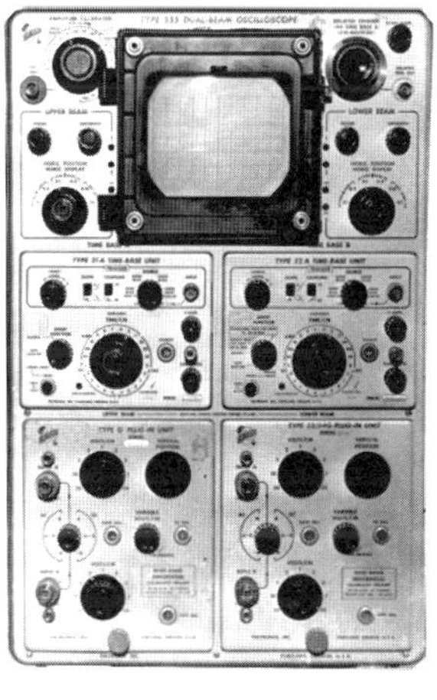

Figure 2-6 - Photograph of the Tektronix 555 dual trace oscilloscope used in recording experimental data.

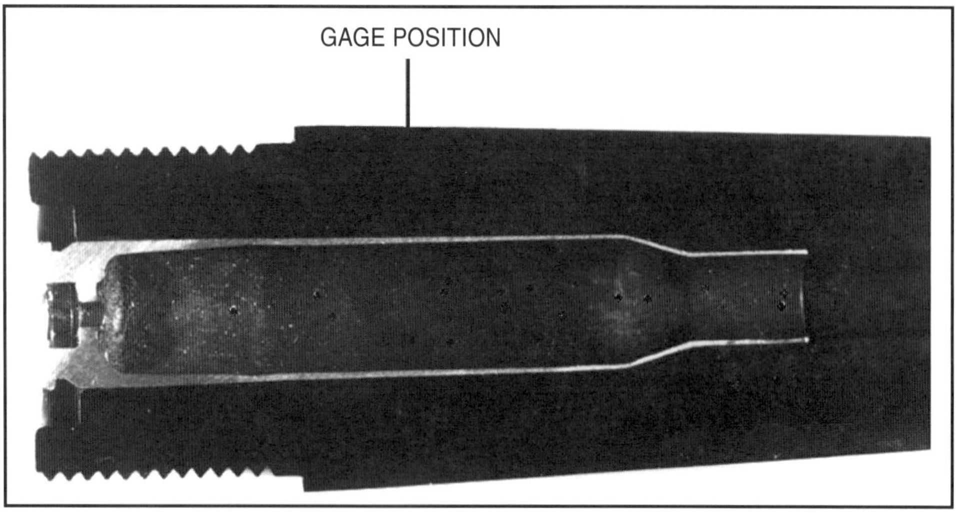

Figure 2-7 - Cross-section of 270 Winchester chamber and cartridge case.

Experimental Chamber Pressure Calibration

Figure 2-7 shows a cross-section of the chamber area of a barrel with a 270 cartridge case, and the longitudinal location of the strain gages is indicated. You can see that radial expansion of the chamber will be restrained at both ends of the chamber with the receiver attached, because the receiver and the neck region are both thicker and restrain the expansion. This is what causes so much trouble in calibration. If the chamber section were longer so that the end restraints didn't have much effect, the gages could be calibrated with a simple experimental test to determine the strain gage amplifier gain. We are going to calibrate the gages by pressurizing a modified cartridge case inside the chamber and noting the scope deflection while we measure the pressure in the case with an accurate high pressure dial gage. The cartridge case is attached to a 1/4 inch outside diameter (OD) steel tube with a 0.15 inch inside diameter (ID). Figure 2-8 shows a photo of the case and 1/4 inch tube,

Figure 2-8 - Photograph of the modified 270 case with a 1/4 inch OD steel tube attached for measuring pressure during calibration. The 1/4 inch tube extends down the bore and out the muzzle where it is attached to the hydraulic cylinder shown in Figure 2-10.

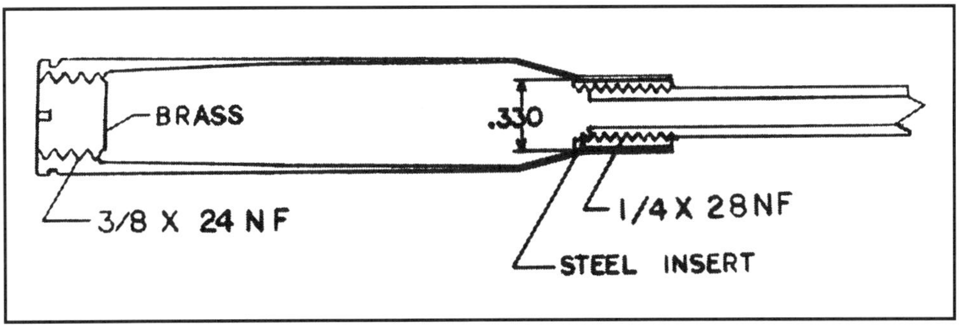

Figure 2-9 - Cross-section drawing of the modified case for calibration showing the case-tube attachment. The left end of the steel insert is larger than the inside diameter of the chamber neck, which prevents the internal pressure from pushing the insert out of the case. The insert is also soldered to the case neck. The hole in the head of the case with the brass plug is required for assembly.

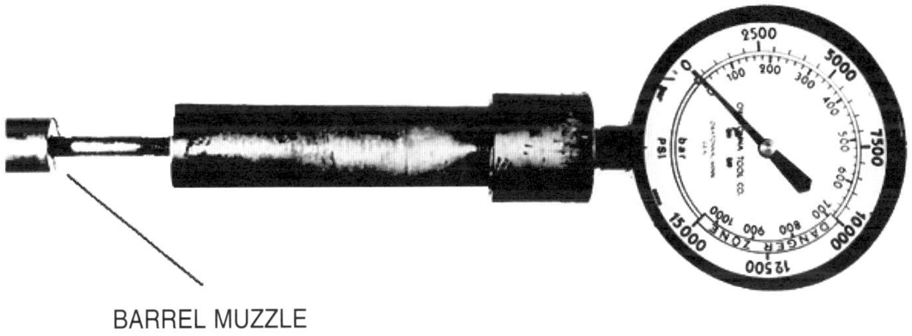

Figure 2-10 - Photograph of Hydraulic cylinder and pressure dial gage used in chamber pressure calibration.

and Figure 2-9 is a cross-section drawing of the modified case. Of course, the tube extends down the bore and out the muzzle where it is threaded into the hydraulic chamber with the 15,000 pounds/square inch (psi) gage (Figure 2-10). The cylinder, tube, and case is filled with hydraulic fluid and as much air as possible is removed. The cylinder is then sealed creating a closed system. When the cylinder is pressurized, a direct calibration of the pressure versus scope deflection is obtained, and we have a pressure measurement that we can be sure of. This is not true of other approaches as other investigators have reported. This method can be used to at least 20,000 psi, however, 20,000 psi gages are expensive, so I chose to use a 15,000 psi gage. The one assumption involved is that the strain gages are linear, which means that we are assuming that the gage response is proportional at the higher strain and pressure levels. Well, strain gages are known to be exceptionally linear, well within our ability to read scope trace deflection and the pressure from the dial gage, so not to worry.

DANGER, do not conduct this experiment without adequate protection to the operator. A hydraulic jet at 15,000 psi is very dangerous, so enclose the experiment in a box. Do not use water as a working fluid, because it could flash vaporize if the cylinder fails, causing a violent explosion. Use a heat treatable steel for the cylinder, such as 41L42 heat treated to at least 120,000 psi. Mild steel is too weak.

It would also be a good idea to decrease the ID of the 1/4 inch tube from 0.15 to 0.125 inch to improve the strength in the region of the threads on each end of the tube. Loctite was used as a seal on the cartridge joints and Teflon tape on the other joints. A new cartridge case was used, because it is softer than a case that has been fired and resized, and will conform to the chamber more quickly.

The resulting calibration is shown in Figure 2-11, where the scope vertical deflection in cm (.05 volts/cm) is plotted versus the pressure measured by the dial gage. The points do not lie on the straight line until a pressure of about 12,000 psi is reached, then the last three points are on the straight line through the origin. Lames' equation can be used to show that a pressure of 12,000 psi is required to expand the head region of the case so that it is in hard contact with the inner wall of the chamber. Once the case has expanded sufficiently so that it is in hard contact with the chamber, the expansion of the chamber

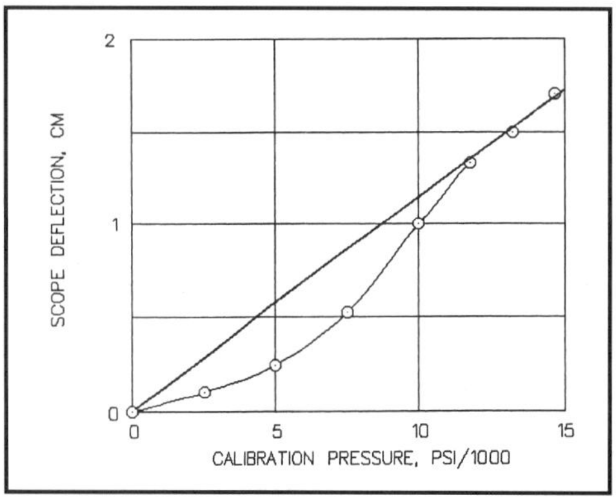

Figure 2-11 - Chamber pressure experimental calibration. At 12,000 psi the case has expanded so that it is in contact with the chamber. At 12,000 psi and above the calibration curve is a straight line (linear).

becomes proportional to the pressure and a straight line calibration results. This results in a voltage output of 0.285 volts (2.85 cm scope deflection) at a chamber pressure of 50,000 psi and a strain gage bridge supply voltage of 6.00 volts. By repeating the experiment several times the reading accuracy of the data approaches a few tenths of a percent. The pressure dial gage accuracy is quoted by the manufacturer to be 1-1.5%. Consequently, the accuracy of the calibration is probably in the range of 1 to 2%, which amounts to a variation of 500-1000 psi at a chamber pressure level of 50,000 psi. This accuracy is more than adequate for our purposes and is likely more accurate than much of the published chamber pressure data.

Theoretical Calibration

There is a theoretical method of calibration, which was described by Brownell in 1965 (Reference 1), that uses Lames' equation for thick-walled cylinders to calculate the circumferential strain at the outside surface. This strain value can be used with the known electronic characteristics of the strain gages and amplifier to arrive at a theoretical calibration. Unfortunately, Lames' equation is based on the assumption that the thick-walled cylinder has an infinite length and constant diameter. In the real case we have a short thick-walled cylinder that is reinforced on one end by the forward receiver ring and on the other end by the tapering chamber. As a result, Lames' equation overestimates the amount of strain by roughly 20%, because the chamber section of the barrel is considerably stronger than a uniform thick-walled cylinder. This means that at a true peak chamber pressure of 50,000 psi the theoretical calibration would indicate a pressure of only 40,000 psi. Consequently, I much prefer the experimental calibration approach.

Strain Gage Electronics

Since someone may wish to use this method to measure chamber pressure, the electronic equipment and method will be briefly described. The strain gages are Model number HBM 6/120LY11 gages purchased from Omega Engineering, Inc. (1-(203)-359-1660) and are connected in a bridge as is shown in Figure 2-4. The bridge supply voltage is furnished by four 6 volt lantern batteries connected in parallel and is monitored with a digital voltmeter accurate to 0.2%. The cases of the batteries are connected to the common ground. Any variation in this voltage directly effects the calibration in a proportional manner. The 120 ohm resistors should be as precise as possible. The leads are unshielded twisted pairs. A photo of a strain gage amplifier

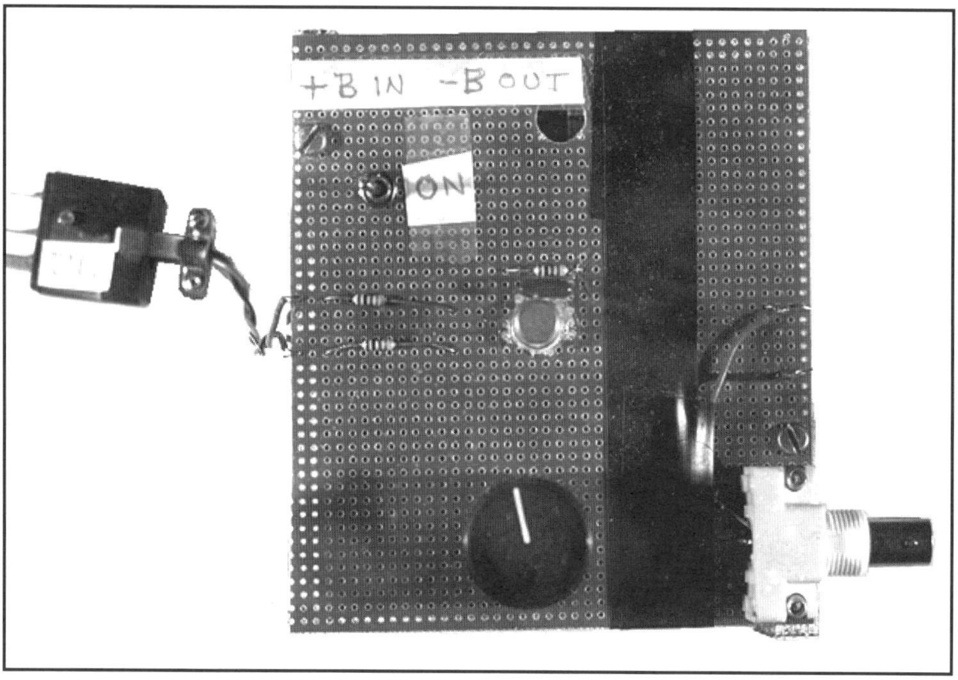

Figure 2-12 - Photograph of the strain gage bridge amplifier.

is shown in Figure 2-12 and the circuit for the amplifier is shown in Figure 2-13. The gain depends on the precision of the input resistors and the feedback resistor and is constant to about 5 kc, which is adequate for this application. The LM101A op-amp chips are cheap, easy to use and stable, and

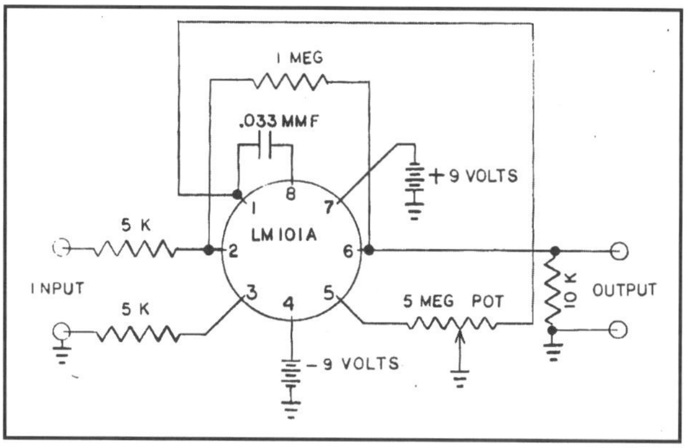

Figure 2-13 - Circuit diagram of strain gage bridge amplifier.

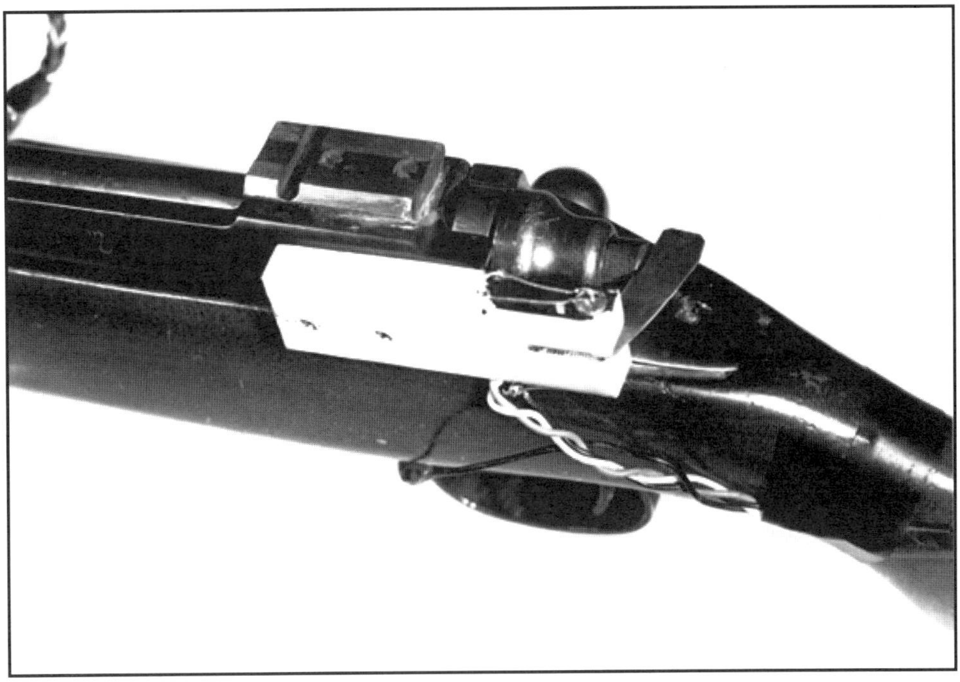

Figure 2-14 - Photograph of the scope trigger switch that starts the instrumentation.

are made by several companies. The positive and negative 9 volts required by the amplifier is supplied by 9 volt batteries.

The output is adjusted to zero with the potentiometer, using a digital meter on the output. The output of the amplifier is connected to a Tektronics 555 dual trace oscilloscope through 12 foot shielded cables. The 1 kw 120 volt AC for the whole thing is supplied by a small portable generator.

The switch that supplies a 9 volt pulse to trigger the scope sweep is shown in Figure 2-14. A brass plate was soft soldered to the firing pin cocking piece, which interrupts an infrared light beam that shines on a photodiode. Both the light source and the receptor are enclosed in the plastic block shown bolted to the rear receiver ring. The infrared light source is a light emitting diode (Radio Shack No. 276-066A) that is powered by 1.5 volt batteries. The phototransistor (Radio Shack No. 276-145) is powered by a 9 volt battery, and requires no external circuitry. The white plastic block is usually covered with black tape to reduce stray light. The scope has an adjustable time delay that is used to move the pressure time history to an acceptable location on the scope face relative to time.

The noise in the strain gage circuitry is low, about 0.2 mv. This results from the low impedance of the bridge (120 ohms) connected to the higher input impedance of the amplifier (5,000 ohms) and the relatively low output impedance of the amplifier (10,000 ohms) connected to the high input impedance of the oscilloscope (10 megohms). This has the effect of looking like a shorted input to the amplifier and the oscilloscope as far as noise is concerned.

Results

Now that we have suffered through all this background information—I warned you that you would find out more about chamber pressure measurement than you wanted to know—let's look at the results. Five loads of IMR4831 (53,55,57,59,61 grains) were tested in Remington 270 cases with 90 grain bullets. The scope trace is shown in Figure 2-15 for the 57 grain load. The upper traces are the chamber pressure and the lower traces are one channel of the receiver ring moment, which we will ignore for the time being. The major divisions on both scales are in centimeters and time is on the horizon-

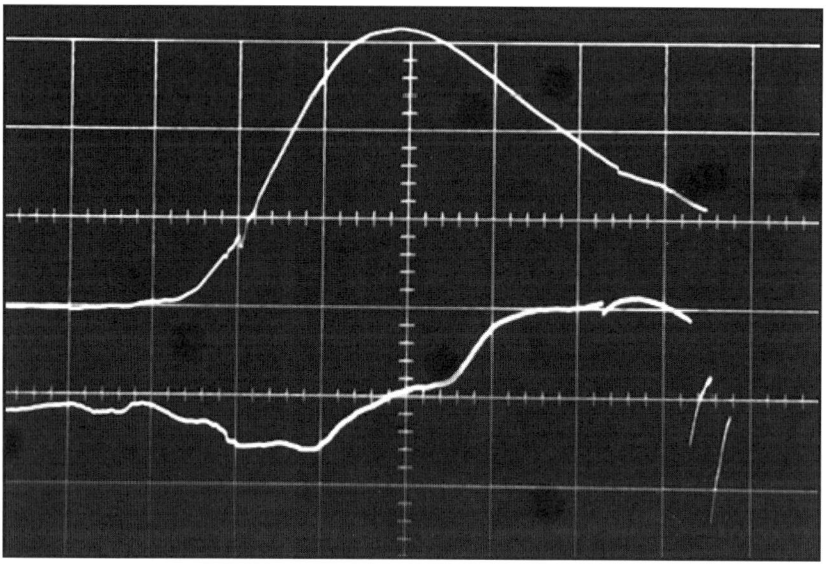

Figure 2-15 - Scope trace of chamber pressure versus time for a load of 57 grains of IMR4831 and a 90 grain 270 bullet. The major divisions are 1 centimeter (cm). Horizontal scale is set at 0.2 msec/cm.

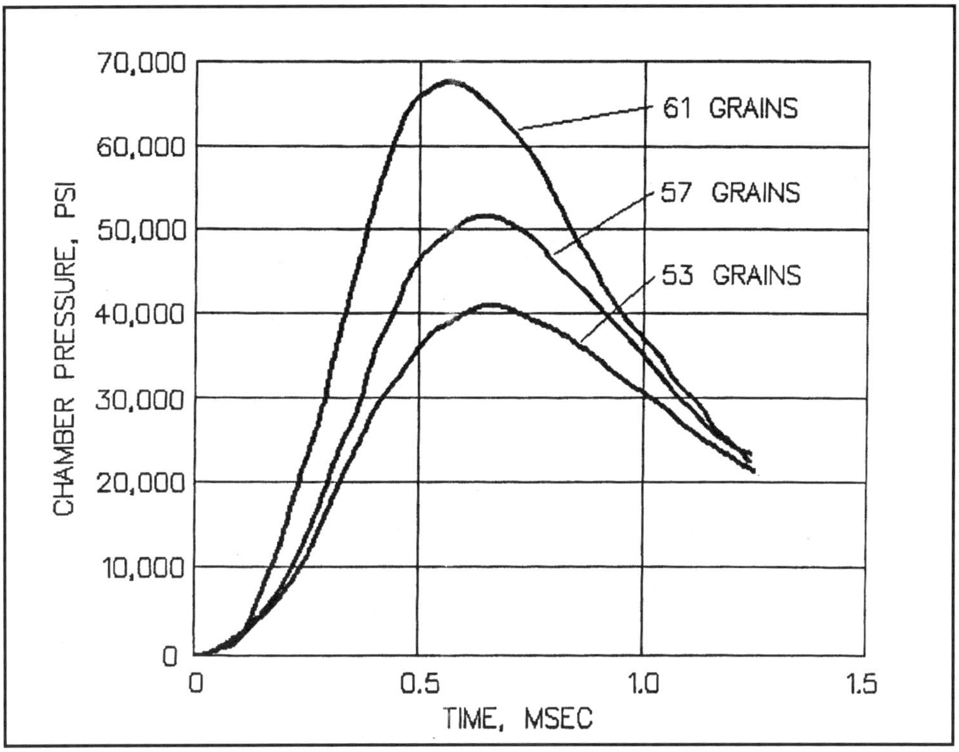

Figure 2-16 - Experimental chamber pressures obtained for three different powder loads in the Winchester 270 cartridge, using Remington cases, 90 grain Sierra bullets, and IMR4831 powder. The oscilloscope traces are terminated when the bullet exits the muzzle.

tal scale reading from left to right at 0.2 msec. The traces are terminated, that is deflected off scale momentarily by a wire taped to the muzzle, which adds a 9 volt signal when the bullet emerges from the muzzle and contacts the wire. We need the bullet exit time when we compare the velocity measured by the chronograph to that obtained from the experimental pressure measurement. Incidentally, the 59 grain load fills the case to the base of the bullet and the 61 grain load completely fills the case and requires about 0.15 inch compression by the bullet when the bullet is seated. The pressures obtained by reading the oscilloscope traces and using the experimental calibration are shown in Figure 2-16 for three powder loads. Notice that these curves are very regular and orderly, which is what we should expect. Several measurements were made which allows us to determine the variation in peak pressure and velocity with different powder loads. The data are shown in the following table.

TABLE 2

270 Winchester Case, IMR4831 Powder, 90 Grain Sierra Bullet

Powder Grains	Average Peak Pressure	Peak Pressure Variation ±%	Average Velocity	Velocity Variation ±%
53	40046	2.3	2758	6.0
55	48709	1.0	3002	3.1
57	53580	3.3	3018	1.3
59	64924	—	3154	<0.6
61	69366	2.5	3235	<0.6

There are several things of importance to be noticed. First, the variation of the peak pressure about the mean is roughly 2% (1000 psi) while the velocity variation about the mean is much bigger with light loads and as the case is filled more completely with powder, the velocity variation reduces to <0.6%, which is the resolution of the older chronograph that was used (see Chapter 10). In other words the velocity variation is constant at the two top loads as far as the chronograph can tell (Figure 2-17). The symbol < means less-than.

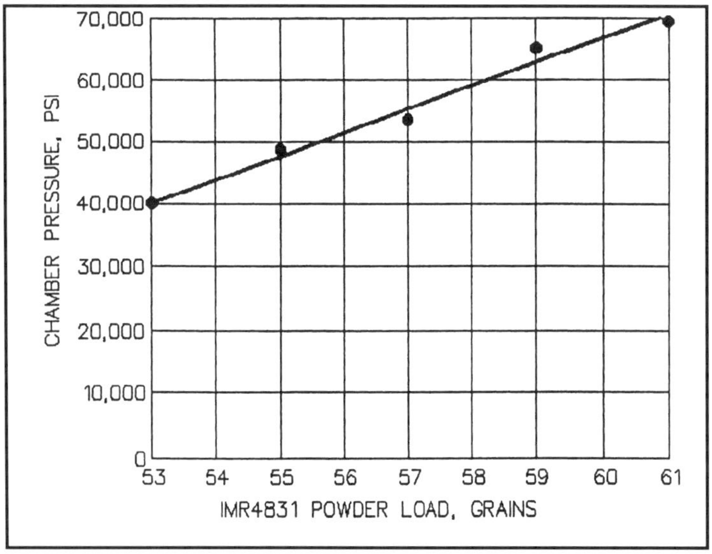

Figure 2-17 - Peak chamber pressure for various IMR4831 powder loads and 90 grain 270 bullet.

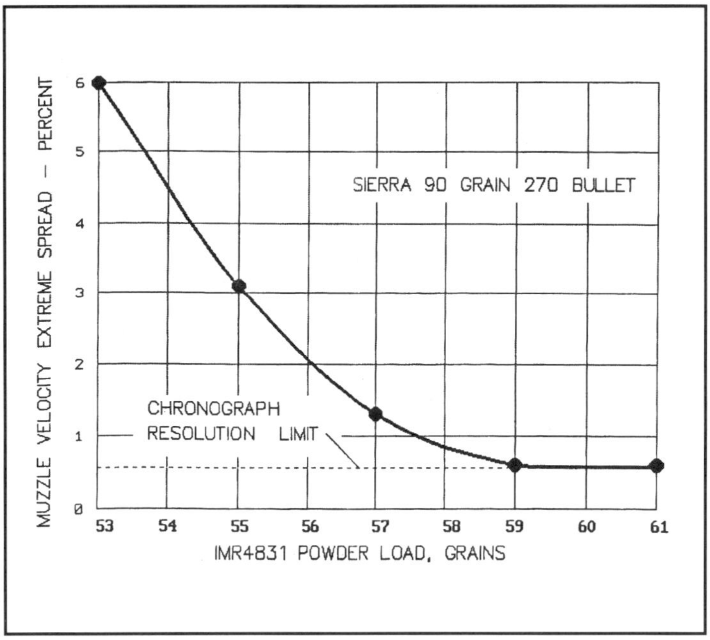

Figure 2-18 - Variation of muzzle velocity extreme spread in percent of velocity for various IMR4831 powder loads with 90 grain 270 bullets.

Graphs of the peak pressure and the velocity variation are shown in Figures 2-17 and 2-18. This data tells me that the pundits are right in that the case has to be filled to a level near the base of the bullet for consistent velocities, which is important in achieving good accuracy. However I have never seen the data before to support this conclusion. The usual explanation is that the powder distributes itself differently when there is free space in the case, leading to variations in powder ignition. In addition, the data tells me that an optimum load is probably in the neighborhood of 57 grains. A load of 58 grains will give a peak pressure of about 60,000 psi, and a calculated peak tensile stress in the chamber of about 73,000 psi. This is about 40% of the yield stress of the barrel steel, neglecting stress concentrations (sharp

Figure 2-19 - Photograph of case heads showing change in primer appearance for increasing chamber pressure from left to right.

corners, etc.) which are always present to some degree. Therefore, as far as this citizen is concerned, 58 grains would be the maximum load behind a 90 grain bullet in this rifle. And this rifle may have lower pressures than most commercial rifles, because it has a throat half cone angle (0.75°) about half that of standard 270 chambers (1.5°). This was done with a special reamer to purposely ease the engraving process in order to reduce bullet distortion (more about this later). I have read numerous times that full case loads of 4831 are safe in a 270 Winchester. A full case load of IMR4831 behind a 130 grain bullet would be exceedingly dangerous—so much for the pundits that guess without having the facts!

I have also read that you can estimate the pressure by the condition of the primer. This happens to be right, but it is crude and requires a calibrated eyeball. To help you calibrate your eyeball, a photo of the heads of five cases fired with the five loads shown in Table 2, is shown in Figure 2-19. The peak chamber pressure increases from 40,000 psi on the left to about 70,000 psi on the right. You can see that the edges of the primer (Federal No. 210) and the edges of the firing pin indentation get sharper and more square with increasing pressure. This depends on the primer, but with experience it can be a rough indication of pressure level. I do not believe that firing pin indentation cratering means much, because it depends too much on the shape of the nose of the firing pin and the clearance between the pin and the hole in the bolt face. This type of primer cratering is more likely to occur on poorly made rifles, such as military rifles manufactured with excessive clearances. Figure 2-20 shows a situation where the primer tells a great deal—you see it's missing!!! The pressure must have been enormous, because the primer pocket

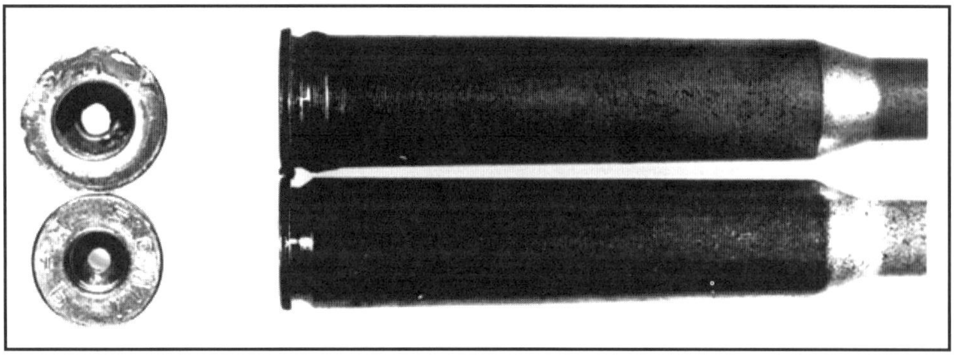

Figure 2-20 - Photograph of a 270 case head subjected to excessive chamber pressure (top) compared to a normal case (bottom).

was expanded from 0.210 to 0.275 inches. Compare the deformed case in the top of the photo with the standard 270 case in the bottom of the photo. You see, this fellow accidentally got a 7mm (0.284 inch diameter) bullet mixed with his 270 (0.277 inch diameter) bullets and tried to shoot a 7mm bullet through a 270 barrel. It didn't seem possible to get this combination in the chamber, until I made a measurement of the chamber neck diameter and found out it was possible. The shooter was lucky, because the gun had a bolt head that completely enclosed the case head and it didn't explode, but the gun was completely ruined. Anybody can make a mistake, so be careful when you handload.

Calculated Interior Ballistics

In order to get the velocity, bolt force, pressure on the base of the bullet, and distance the bullet moves in the barrel, we have to resort to the use of a Personal Computer (PC) and a special internal ballistics computer code that I developed. We have to do this, because we need this information to isolate and explain the various root causes of inaccuracy. We also need to go through these calculations to further validate the chamber pressure measurement. This may seem to be an overkill as far as the chamber pressure measurement is concerned, but this gun nut has to be certain. Fortunately, it's not very complicated, so hang in there.

There are two types of pressure, static pressure and dynamic pressure. Dynamic pressure (q) results from the motion of the gas and is

$$q = (1/2)*Rho*V^2$$

where Rho is the gas density and V is the velocity. Static pressure is the pressure exerted by stagnant gas. Total pressure is the sum of the two and is the pressure you feel when you hold your hand out the window of a speeding car. A long time ago a fellow named Bernoulli found out that as long as the gas flow velocity doesn't exceed the speed of sound (subsonic), the total pressure is constant and is the sum of the static and dynamic pressures. What this amounts to is that the chamber pressure is equal to the total pressure or the stagnation pressure, because the gas velocity is negligible in the chamber. The static pressure on the base of the bullet is equal to the total pressure minus the dynamic pressure. This means that the static pressure on the base of the bullet, which drives the bullet down the barrel, decreases as the bullet picks up speed. Therefore,

$$Pb = Pc - (1/2)*Rho*V^2$$

where Pb is the static pressure acting on the base of the bullet and Pc is the measured chamber pressure. The gas density Rho can be obtained from the equation of state for a gas, which is

$$P = R*Rho*Tb$$

where R is a constant (1716 ft^2/sec^2/deg R) and Tb is the effective gas temperature (powder burn temperature) in degrees Rankine (degrees F + 459). We know that the temperature is about 6000°F or about 6500°R from reference data. Consequently, we can get the pressure acting on the base of the bullet, subject only to the accuracy with which we know the gas temperature. The other thing that we don't know about the internal ballistics is just how much of the powder weight is accelerated with the bullet. We know from theory that about half the powder weight is effectively accelerated with the bullet if the powder is completely burned before the bullet exits the muzzle. It will turn out that IMR4831 powder is so slow burning that it doesn't all burn out before muzzle exit of the bullet and so about 60% of the powder is effectively accelerated at the speed of the bullet. Now this makes two things that we don't know precisely; powder burn temperature and powder acceleration fraction. However these can be determined with precision, by trial and error, because we know the muzzle velocity and peak chamber pressure at two extreme conditions, 53 and 61 grains of powder. In mathematics language, this is the same thing as having two equations with two unknowns, a problem that is easily solved. This is done by making small changes in the gas temperature and powder mass fraction until the calculated peak chamber pressure and muzzle velocity agree with the measured values.

However, before the computer calculations were made two relatively small refinements were made to the computer code. The first is the correction of powder burn temperature with increasing pressure. According to Reference 2 and 3, the powder burn temperature is decreased by about 7% when the pressure increases from 1000 psi to 53,000 psi. The second correction, and the largest, is caused by the fact that the high pressure jet, that squirts out of the muzzle after the bullet exits, continues to accelerate the bullet for about 15 bullet diameters (calibers). Fortunately, the U.S. Army Ballistic Research

Laboratories have been investigating muzzle blast effects for several years (Reference 4 through 8). Gion showed experimentally with an M-16 that it takes about 15 calibers for the bullet to outrun the jet after it leaves the muzzle. All of this is included in the computer code. Later on in Chapter 7 we show the same result. The effect of friction on the bullet is shown to be about 2% in Reference 9 and was ignored in the computer code because its effect is negligible. The following Table shows the results of the calculations for three loads (53, 57 and 61 grains).

TABLE 3

Comparison Of Measured To Calculated Results From The Internal Ballistics Code

Powder Load Grains 4831	Measured Muzzle Velocity, fps	Calculated Muzzle Velocity, fps	Measured Barrel Time, ms	Calculated Barrel Time, ms
53	2758	2753	1.58	1.47
57	2940	2951	1.35	1.37
61	3235	3239	1.25	1.23

The velocities are in excellent agreement and the time in the barrel data are in fair agreement. The time in the barrel is difficult to measure, because it is difficult to know exactly where to start measuring the time on the pressure traces. Now we know for sure that the measured chamber pressures are correct because we used the measured chamber pressures to derive the calculated velocity. This makes it possible to obtain several calculated parameters that are important.

The experimental chamber pressure and the calculated pressure acting on the base of the bullet for the 57 grain load are shown in Figure 2-21, where you can see that the bullet base pressure is considerably less than the chamber pressure. The bullet simply starts to outrun the gas, and this is one reason that your basic gun is limited in velocity. Figures 2-22 and 2-23 show the velocity and distance the bullet has traveled. Notice that at 0.4 ms the bullet has traveled only 3/8 inch, but the pressure on the base of the bullet (Figure 2-21) is about 35,000 psi and the force on the base of the bullet (Figure 2-24) is about 2,000 pounds. This is more than enough to deform the bullet slightly and make it expand diametrically. This was proven experimentally by others by firing bullets from a very short barrel (6 inches) and recovering the bullets. This would greatly exaggerate the effect and in fact, these tests showed the diametrical growth was perhaps 10% or more. Unfortunately, no one knows how to measure this effect in a normal length barrel where the expansion would be much less. It could be calculated with a very complicated finite element computer code, but this is beyond the scope of this work. Bullet distortion causes inaccuracy, because it can result in a canted base that causes dispersion by interacting with the muzzle blast, causing the bullet to be deflected (Chapter 7). Bullet distortion can also cause an offset center of gravity (cg), which results in a tangential velocity coming from the high rate of spin (Figure 2-25) imparted by the rifling, and this also causes dispersion (Chapter 9). We will show these effects later in great detail, both experimentally and theoretically.

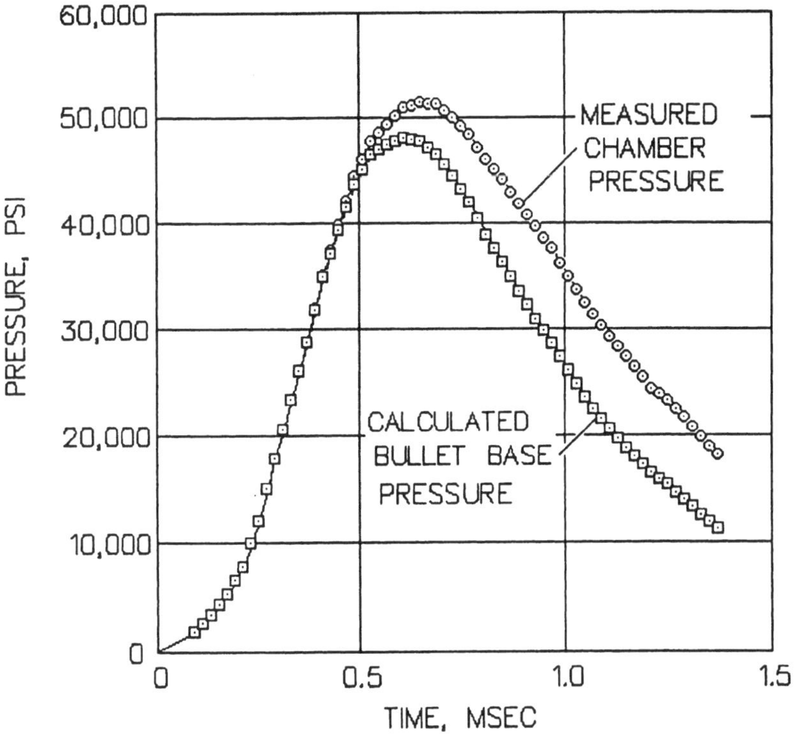

Figure 2-21 - Comparison of chamber pressure to the pressure acting on the base of the 270 bullet.

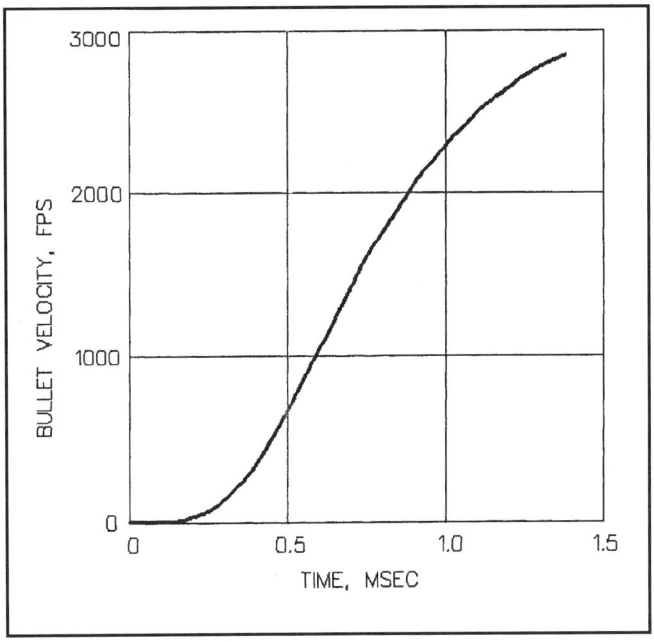

Figure 2-22 - Bullet velocity versus time in bore. The velocity is obtained by integrating the measured pressure for the 57 grain load shown in Figure 2-17.

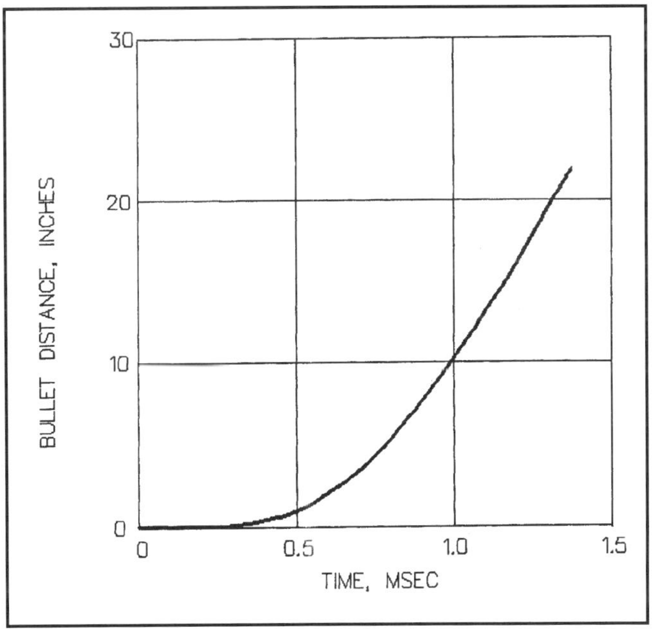

Figure 2-23 - Distance bullet has traveled down the barrel versus time in bore. The distance is obtained from an internal ballistics computer code that integrates the velocity.

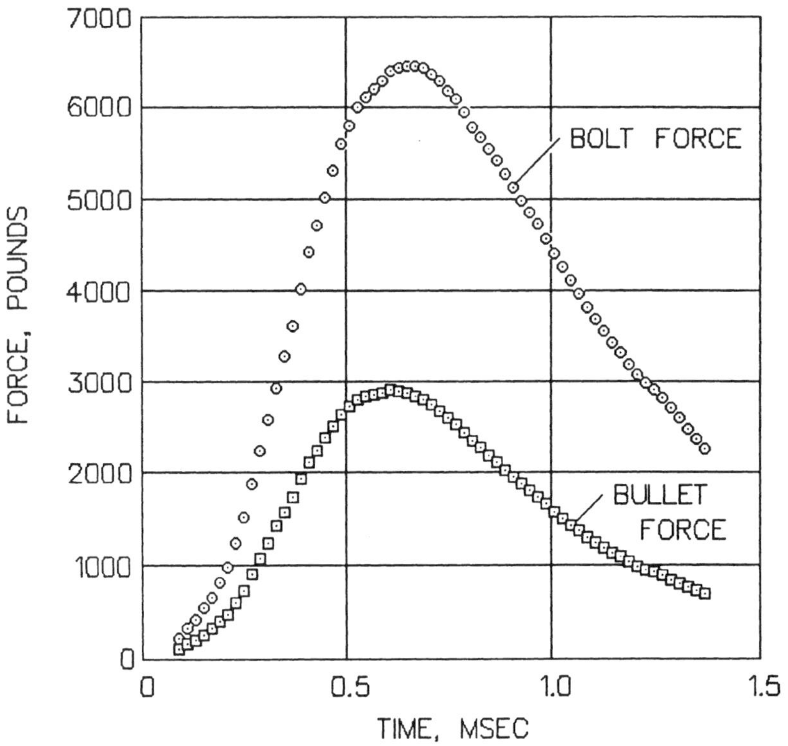

Figure 2-24 - Forces acting on the bolt face and on the base of the 270 bullet versus time in bore. Forces were calculated using the measured chamber pressure for a 55 grain load of IMR4831.

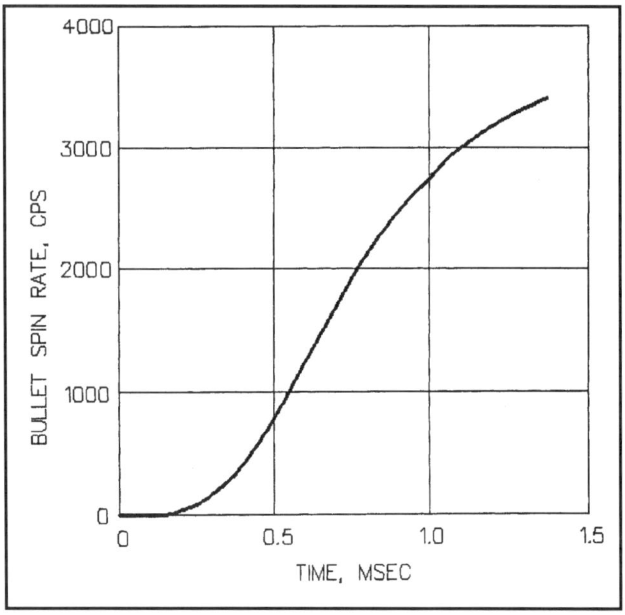

Figure 2-25 - Bullet spin rate in cycles per second (cps) versus time in bore for a 10 inch twist. Note that the bullet is spinning at about 3400 cps (204,000 revolutions per minute) at muzzle exit.

Bullet Engraving Force

A sizable force is required to push the bullet into and through the throat. I decided to try to measure the engraving force by pushing a 270 bullet through a throat with a calibrated hydraulic press. The measured engraving force was 1200 pounds. However one has to reduce this by roughly a factor of two to account for the difference between static and sliding friction force. The measurement is made under the condition of static friction and sliding friction is about half that of static friction forces. Consequently, we end up with an engraving force of about 600 pounds. This translates to a pressure on the base of the bullet of about 10,000 psi. The chamber pressure data in Reference 1 indicates that a minimum chamber pressure of about 10,000 psi (745 pounds force) is required to push a 30 caliber 150 grain bullet through a barrel. My guess is that the force is proportional to the caliber and the pressure required for engraving in the throat remains more or less constant at about 10,000 psi. If you compare bullet motion in Figure 2-23 with the pressure in Figure 2-21 you will see that the bullet doesn't move until a chamber pressure of about 10,000 psi is reached. Once the bullet has passed through the throat the friction force required to move the bullet in the bore is about 80% (480 pounds) of the engraving force, although this friction force is likely reduced as the bullet moves faster.

The pressure change caused by changing bullet seating depth in the case can also be deduced from Reference 1 for 0.308 caliber bullets. The peak chamber pressure will drop about 1000 psi for every 30 mils of additional distance (free run) between the bullet and contact with the lands in the throat. In other words, if you seat the bullet so that it has about 60 mils of free run before contacting the lands, the peak chamber pressure will be reduced by about 2000 psi. This means that the chamber pressure is not very sensitive to seating depth. I like to use somewhere between 5 and 30 mils free run for heavy 270 or 30 caliber bullets and I like to seat light 6mm bench rest bullets in contact with the lands. Most bench rest shooters seat their bullets into the lands about 10 mils. This is probably more than you want to know about internal ballistics, but I thought it was interesting.

Bullet Weight Variation

Target shooters often weigh bullets and segregate them according to weight, because variations in bullet weight can cause a variation in muzzle velocity and bullet drop. A variation in bullet gravity drop will cause vertical dispersion. The problem is that no one seems to know how much effect bullet weight variation has on accuracy (vertical dispersion). Well we can calculate the effect of variation of bullet weight on muzzle velocity and measure it experimentally. Calculations with the internal ballistics code show that the fractional change in muzzle velocity for a given fractional change in bullet weight can be calculated from

$$(\delta V/V) = 0.24 * (\delta W/W)$$

where

δV = change in muzzle velocity

V = muzzle velocity

δW = change in bullet weight

W = bullet weight

From this equation we can see that a variation in bullet weight of one percent will result in only a 0.24 percent change in muzzle velocity. The reason that the muzzle velocity doesn't change as much as one might expect, is that there is a compensation factor involved. The heavy bullet will cause the peak pressure to be higher and to peak earlier than that of the lighter bullet, thus providing a partially compensating effect. In order to check this theoretical calculation, Walter Jankowski of Cook Bullets made up some 6mm bullets in 65 and 75 grains weight in the same jacket with identical shape. This amounts to a 15.4% increase in weight, and according to the equation we would expect a 3.7% change in muzzle velocity. I test fired these bullets in a 6mm bench rest Heavy Varmint class rifle and got a variation of 3.2% in muzzle velocity using an Oehler 35P chronograph with six foot optical gate spacing. There is more about the effect of bullet weight variation on vertical dispersion in Chapter 10.

Cartridge Case Failure

Since I have used short cartridge cases with excess headspace for experimental purposes, the reader may have come away with the impression that excess headspace is harmless and perhaps even helpful in reducing dispersion. So I want to dispel any notion that excess headspace, meaning more than a few mils, is acceptable. Repeated firing of cartridges with excess headspace will certainly cause case head separation, which is potentially dangerous. Recall that the experimental data were obtained under the condition of remote firing. Also, these short cases were specially treated to help prevent case failure. If you can detect any primer protrusion on a fired case, you have excess headspace and you need to correct it. Figure 2-26 shows four common case failures. The first case on the left is a standard 270 case that has suffered a case head separation as a result of repeated excess resizing. The second case from the left is a 300 magnum case that has a head separation resulting from excessive headspace. The third case from the left is a 30-06 case that has an axial split near the head of the case. This results from repeatedly firing resized cases in an oversize chamber and is probably one of the most dangerous types of failure. Oversize chambers generally occur in military rifles, because they have to operate in dirty conditions and are not intended to be fired with reloads. This particular one came from a surplus 30-06 Springfield that I had some 40 odd years ago. If you have one of these old turkeys, it is best not to use old cases. The fourth and last case from the left is the result of firing a 7mm-08, which is a 308 Winchester (7.62mm NATO) case necked down to 7mm in a 280 Remington chamber which is much like the 30-06 chamber

Figure 2-26 - Photograph of four cartridge cases showing different modes of failure. Head separation is shown on the first two cases from the left. Axial split near the head on the third case from the left. The case on the right resulted from firing a 7mm-08 in a 280 Remington chamber.

which is shown just to the left in the figure. Well the case is about 0.44 inch too short for the chamber and it simply expanded to fit the chamber. Fortunately, the bullet was the right size and nobody got hurt. The 'bottom line' is—be careful.

Summary

We now have all the internal ballistic parameters that we need to help isolate and minimize the other causes of inaccuracy. The force accelerating the bullet shown in Figure 2-24 is the same as the recoil force, the force that pushes the gun to the rear, and will be used later to analyze barrel vibration effects.

CHAPTER 3
CHAMBER AND THROAT DESIGN

While this is not intended to be a gunsmithing book, because there are already several good ones available, we need to describe how the chamber is machined and why it may make a difference. If the chamber or the throat is not concentric with the bore or if it is oversize, the bullet will have to rattle around (balloting) before it can line up with the bore center line and pass through the throat. This can cause the bullet to be deformed in an asymmetric manner. A throat with an overly abrupt taper can also cause bullet deformation. Just how this deformation takes place is a guess, but it most likely due to a very slight amount of in-bore canting. This will cause a center of gravity offset from the bore center line, which can cause dispersion. It also can cause the base to be canted resulting in interaction with the muzzle blast causing dispersion. It could also result in a tilt of the principal axis, which causes an error that will be discussed later. We minimize these effects by cutting the chamber and throat concentric with the bore and we reduce the slope of the throat to about one half that of standard throats.

If you make a cast of the chamber and throat of a commercially made rifle, you will most likely find that they are not concentric with the bore axis—in other words, they are off center. A photo of front and back views of a chamber cast of a factory chamber is shown in Figure 3-1. If you look carefully at the top picture you can see that the rifling starts at the entrance to the throat,

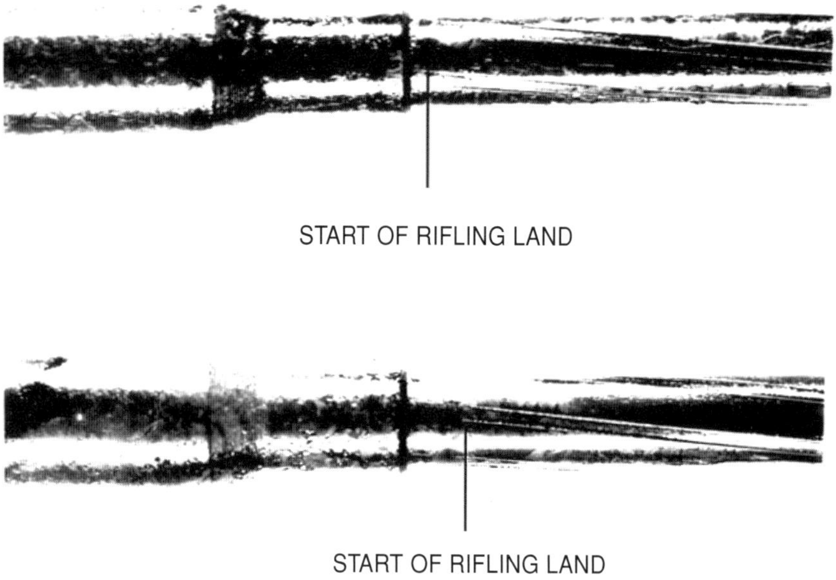

Figure 3-1 - Photograph of a cast of a factory production rifle throat showing offset chamber and throat. The cast has been rotated for the bottom photograph.

while in the bottom photo the rifling starts at about 0.18 inches from the start of the throat. Since the throat half cone angle is about 1.5 degrees, the amount of throat offset from the bore center line can be estimated to be about 2.5 mils (0.0025 in). That's several times the error that should be tolerated. This happens, because chambering is done very quickly in automatic machines at the factory, and requires the use of reamers without pilots. Any experienced machinist will tell you that a pilotless reamer will likely run off center and can greatly enlarge a hole. Figure 3-2 shows a photo of a chambering reamer and a separate throating reamer, where you can see the pilots on the ends of both reamers. The pilot is the short cylindrical section on the end that fits the bore very closely and guides the reamer as it cuts, so that it keeps the

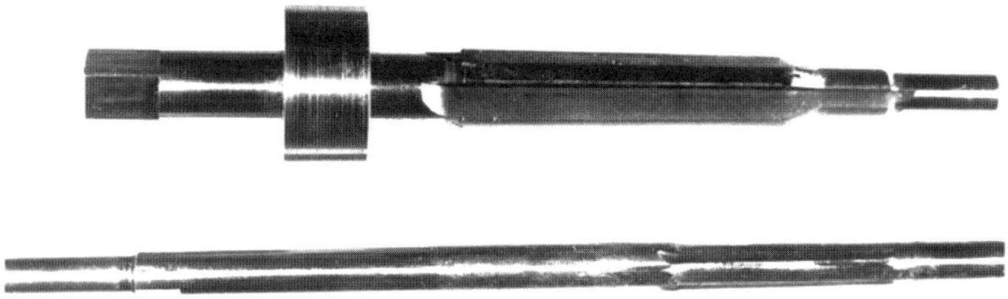

Figure 3-2 - Photograph of separate chambering and throating reamers with pilots.

chamber and throat centered with respect to the bore. The reamers are made of high carbon steel drill rod and are rough machined before hardening. They are oil quenched from 1600°F and tempered at 500°F. Then they are ground with a tool post grinder and sharpened by hand, and they last a long time without resharpening. There are adjustable stops on the shanks to control the final depth of cut. A piloted reamer is difficult to use in factory production, because chips are likely to interfere and cause the pilot to gall and seize. Also, you have to stop and clean the reamer often, about every 1/16 inch of cutting depth. You have to go very slow, operating the lathe at 60 RPM. The chambering operation is shown in Figure 3-3. There is nothing particularly new in this set up, except that it is optimized by the fact that the lathe spindle has a hole through it large enough to accept the barrel. You can do good work with a smaller lathe by holding the muzzle in the head stock and the chamber end in a steady rest, but it is more difficult. It takes me about two hours to machine a chamber and throat, after the barrel is set up in the lathe. This kind of time is much too long for factory production. While I haven't measured a

Figure 3-3 - Photograph showing the chambering operation.

lot of chambers in a lot of guns from different manufacturers, I have yet to see a good one in a factory rifle. However, I am sure that occasionally a pilotless reamer does run on center and produce a more accurate chamber. Also, I haven't inspected any recent factory sporters so it's possible that the factories are doing a better job now, but I doubt it.

An oversize chamber is another problem that shows up in factory rifles. This can result from using pilotless reamers, oversize reamers, or a poorly sharpened reamer. A factory chamber must have enough clearance to accommodate cases with thick necks caused by excessive full length resizing and the build up of carbon and dirt in the chamber, otherwise a dangerous high pressure condition could result. They also have to accommodate a wide variety of commercial ammo. The case neck must be able to expand enough to release the bullet. The chamber body and neck that I use in the 270 have a diametrical clearance of 3 to 4 mils, and I have not had any problem. This is, in effect, a minimum chamber and if you are careless about the make or condition of the hulls that you stuff in your rifle, it could cause big trouble. Bench rest rifles have undersize chamber necks and therefore the case necks must be turned down before loaded rounds will chamber. The Lapua factory 220 Russian cases used for 6PPC bench rest rifles have neck walls that are about 13 mils thick. The case neck is turned down so that the final case neck thickness is about 8.5 mils and the variation in thickness is kept to less than 0.1 mil. The loaded rounds are also carefully measured before firing to make sure they will chamber. The radial clearance between the neck of a loaded round and the chamber neck in a bench rest gun is usually only 0.4 to 0.7 mil (0.0004 to 0.0007 inches). They also seat the bullet into the lands, which helps to center the bullet. However, seating the bullet in the case so that it contacts the rifling in the throat also increases the peak chamber pressure, which is not desirable. Evidently the bench rest shooters have found that having the bullet centered in the bore is important, and I think they are right. It is not uncommon for a chamber in a military barrel to have a radial clearance of 5 mils (0.005 inches) the whole length of the chamber.

Seating depth of the bullet in the case has an effect on just how close to the center the bullet will line up. Obviously, the bullet will be centered if it is in complete contact with the lands, however Reference 1 showed that peak chamber pressure decreases if the bullet has a free run before it contacts the lands. Since a minimum in peak pressure for a given load implies minimum bullet distortion, the author prefers a seating depth that will provide about 0.010 inches into the lands in the case of a bench rest gun with light bullets and about 0.020 inches of bullet free travel before the bullet contacts the lands in the case of a sporter shooting heavy bullets. This means that the maximum bullet offset in a sporter can only be about 0.2 mils (0.0002 in) with a throat cone half angle of 0.75 degrees. Although there is no way to prove it, a bullet offset of as much as a few tenths of a mil is probably small enough to minimize bullet deformation. The sketch in Figure 3-4 demonstrates what we are

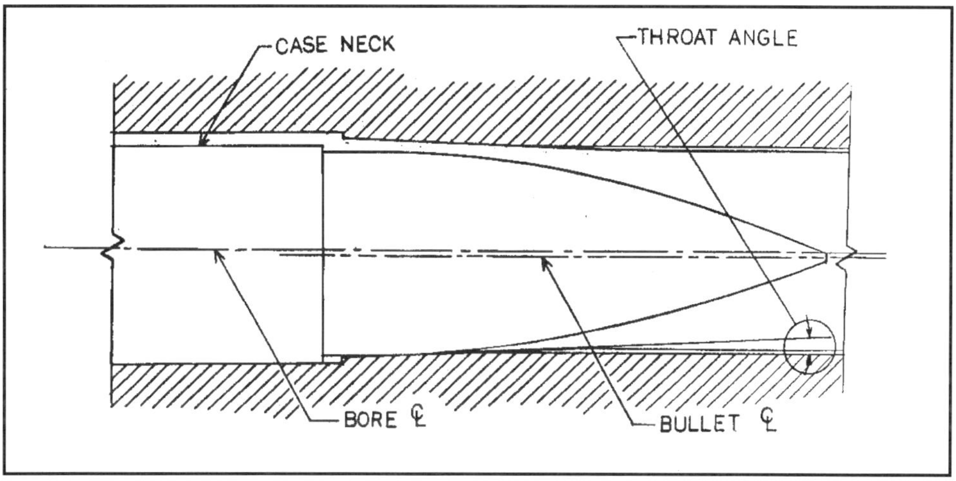

Figure 3-4 - Sketch showing the effect of seating depth on bullet centering.

talking about. This shows a concentric throat and a chamber that is concentric but oversize. If the bullet is moved forward it will also move upward closer to the center line of the bore. It also shows why a shallow throat half angle improves bullet alignment with the bore, while allowing a satisfactory amount of free travel of the bullet before it starts the engraving process in the rifling. I have not tried throat half angles of less than 0.75 degrees, because bullet stripping could occur if the angle is too shallow and I doubt that a more shallow angle would improve the situation. I prefer a steeper slope (2.5° half angle) on the throat of 6mm bench guns because the short bullets seat too far out in the case with a shallow angle throat. Many custom bench rest guns have a half angle of 1.5°.

Standard reloading manuals describe in detail how to set seating dies for proper bullet seating depth. There are special tools available to bench rest shooters who carefully control seating depth. On hunting ammunition it is a good idea to check the final setting to make sure you can't see rifling marks on a new bullet. Otherwise, the bullet may stick in the throat when you unload the rifle and you could have an action full of powder. That is bad news in the field because it is difficult to remove the powder. Obviously, this check should be made with an empty case or at a shooting range with live ammunition.

Making really precision reamers and dies is difficult and takes a lot of time. It helps to make two at the same time, that way you've got a spare to continue with when you spoil the first one. It usually takes me about three days to make a chamber reamer. A lot of what I learned about making reamers and dies came from Frank A. Hemsted, one of the old time great master custom reloading and bullet die makers. You see, he built a set of bullet dies for me back in 1969 and he got curious about what I was up to and came by for a visit. He stayed several days and taught me a lot about machine work. He was around 80 years old at the time and went to the Happy Hunting Grounds a few years later.

The 270 test rifle used in this book has a Douglas Premium barrel with a throat half cone angle of 0.75 degrees, which is about half that of a factory throat (1.53 degrees). This should ease the entry of the bullet into the throat and reduce bullet deformation. The only proof of this is that this rifle starts out being a little more accurate than a normal factory rifle, and the shallow throat angle plus a chamber centered on the bore axis are the main differences. The other thing about a Douglas Premium barrel is that the bore is straight to start with and is not bent to make it straight. Straightening rifle barrels by bending is one of the main causes of thermal drift of the point of impact, because it introduces stress in the metal crystal boundaries, which relax with increasing temperature. Consequently, when you keep shooting the rifle, the barrel gets hot and it tends to return to its original bent configuration, shifting the point of impact. If the manufacturers would just refrain from taking excessively deep machining cuts and stop bending barrels, this problem would probably go away. I have found that thermal drift can be reduced by firing as much as a hundred rounds at a rate that keeps the barrel too hot to touch. I recently fired a new barrel where the point of impact drifted down and to the left about two inches and stabilized after about 20 rounds. This, in effect, is a rapid stress relieving method, because the high temperature combined with the stresses introduced by firing expedites the process. Use light or medium loads, because the high temperature increases chamber pressure. Custom barrel makers, such as Douglas, Shilen, and others, stress relieve their barrel blanks by soaking them in a furnace with an inert atmosphere to prevent scaling at a temperature of 1020 °F. As far as I can determine, there is very little thermal drift in these barrels due to internal stress. Also, an off center bore or ramp front sights that have been brazed to the muzzle can cause thermal distortion. However, stress relieving won't help a barrel with an off center bore or front sight ramp. There will be some drift due to preferential air cooling on the outside of the barrel. Preferential heating results from the bottom of the barrel being protected by the stock and from wind effects. Some unlimited class bench rest guns have an aluminum tube cover over the barrel, which presumably reduces the preferential cooling effect.

If you wish to do your own gunsmithing there are books on the subject that are helpful (Reference 10, 11, 12). But be careful because guns do blow up. Stay away from old military actions because most of them are not very strong. Figure 3-5 shows a Japanese rifle that was "disassembled" by having cases too long for the chamber. The shooter was injured severely but recovered.

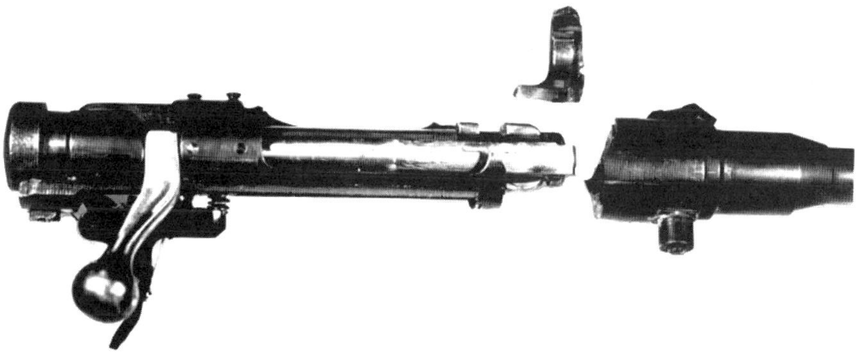

Figure 3-5 - Photograph of a 6.5mm Ariska rifle that exploded as a result of cases being too long for the chamber.

CHAPTER 4
BARREL VIBRATION

Barrel vibration is one of the largest contributors to rifle inaccuracy, however I have been unable to find any evidence of previous experimental work on this subject in the literature. Considerable work has been done on cannon barrel vibration by Ed Schmidt and others at the US Army Ballistic Research Laboratory (References 13, 14, 15, and 16), and the results are somewhat similar to the results obtained in this work on rifles. But the difference in size makes it difficult to apply the cannon barrel work. The only explanation that one can come up with for the lack of work on this subject, as far as rifles are concerned, is that it is a very difficult technical problem that would require the effort of a large research laboratory—and that means a large budget. One does occasionally see an article where someone discusses thc problcm in general terms without any factual data and often reaches an erroneous conclusion. I recall one article where the writer claimed that the stepped configuration of the military Mauser rifle barrel was done to control or reduce barrel vibration! There is absolutely no technical reason to support such a contention and the stepped configuration was undoubtedly used to expedite production machining. Well, in this work we are going to try to find out just how and why the barrel vibrates, and correct the causes of vibration. We will do this by measuring the moment acting on the forward receiver ring which causes the barrel to vibrate and then we will make corrections to the rifle that remove the forces that cause this driving moment. At the same time we will measure the barrel muzzle vibration in the vertical plane with an accelerometer just to make sure that the vertical vibration of the muzzle is reduced. We will also use a barrel vibration computer simulation code as a guide in the design of the instrumentation and to evaluate the effect of barrel vibration on accuracy. While the data presented are restricted to the vertical plane, bear in mind that similar vibration at smaller amplitudes occurs in the horizontal plane due to the same driving moments or forces.

In the course of this investigation we will find that the moments that cause barrel vibration result from the recoil force acting on the recoil lug, and from the bolt thrust acting on the bolt lugs with uneven engagement, and from forces generated by the cartridge case acting on a receiver that is structurally unsymmetrical. We will first eliminate the recoil lug moment with a special bedding device and then reduce the structural asymmetries in the action, resulting in a large reduction in the forces and moments that cause the barrel to vibrate.

All the work on the standard and modified rifle will be conducted on a Winchester 270 cartridge but is applicable to any sporter. At the end of the chapter the 6BR, 6mm Remington and the 6PPC is involved.

Now, a note to the reader: this chapter is one of the longest and most complex in the book. However, every effort will be made to explain everything in physical terms so that the information should be clear to the reader without a technical background. So, hang in there!

Receiver Ring Moment

Barrel vibration is directly related to the moment in the forward receiver ring. Moment is nothing more than the amount of force applied multiplied times the distance or moment arm. For instance, if you apply a one pound vertical force at the muzzle (24 inch moment arm) a 24 inch-pound moment will result at the forward receiver ring, which can be measured with strain gages. In fact, that is how the gages are calibrated. You do it by applying a known force at a known distance, both vertically and horizontally, and reading the amount of oscilloscope deflection. In this case oscilloscope trace deflection sensitivity turns out to be 240 inch-pounds/centimeter with the oscilloscope sensitivity set at 0.05 volts/centimeter. Figure 4-1 shows a rear view of the receiver ring with the location of the four strain gages, and you can see

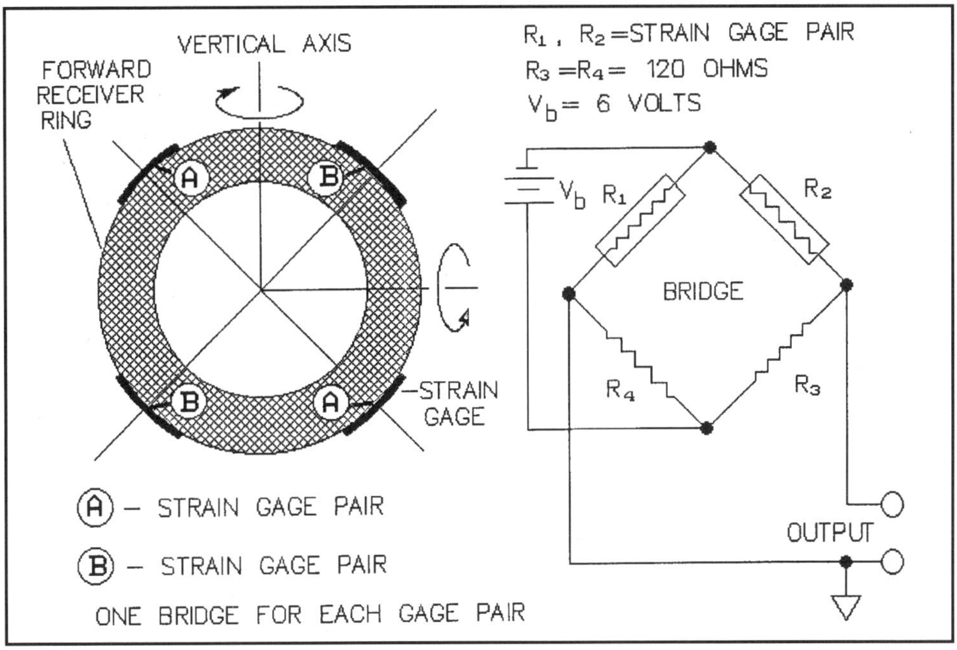

Figure 4-1 - Sketch showing placement of strain gages on the front receiver ring and the strain gage bridge for measuring receiver ring moment.

that they are in two pairs. Each of these strain gage pairs is connected in a bridge circuit also shown in Figure 4-1. A strain gage pair connected in this manner measures the difference in strain between the two gages, which is proportional to the moment. Notice the difference in the circuits of the moment measuring bridge and the pressure measuring bridge (Figure 2-5). The direction of positive moment is shown by the direction of the circular arrows about the axes. The strain gage pairs are rotated 45 degrees to clear the scope mount block and the forward guard screw. However, the outputs from the two strain gage bridges can be combined by the oscilloscope amplifiers to obtain the moment about the horizontal and vertical axes. The moment in the vertical plane, which is the moment about the horizontal axis, is equal to the sum of the A bridge and B bridge moments divided by the square root of two. The moment in the horizontal plane is the A moment minus the B moment divided by the square root of two. The oscilloscope sensitivity in the vertical and horizontal planes is 240 inch pounds per centimeter. The vertical moment is displayed on the upper trace and an upward displacement of the scope trace represents a moment that would push the muzzle upward. The horizontal moment is displayed on the lower trace and an upward displacement of the trace is equivalent to moving the muzzle to the left.

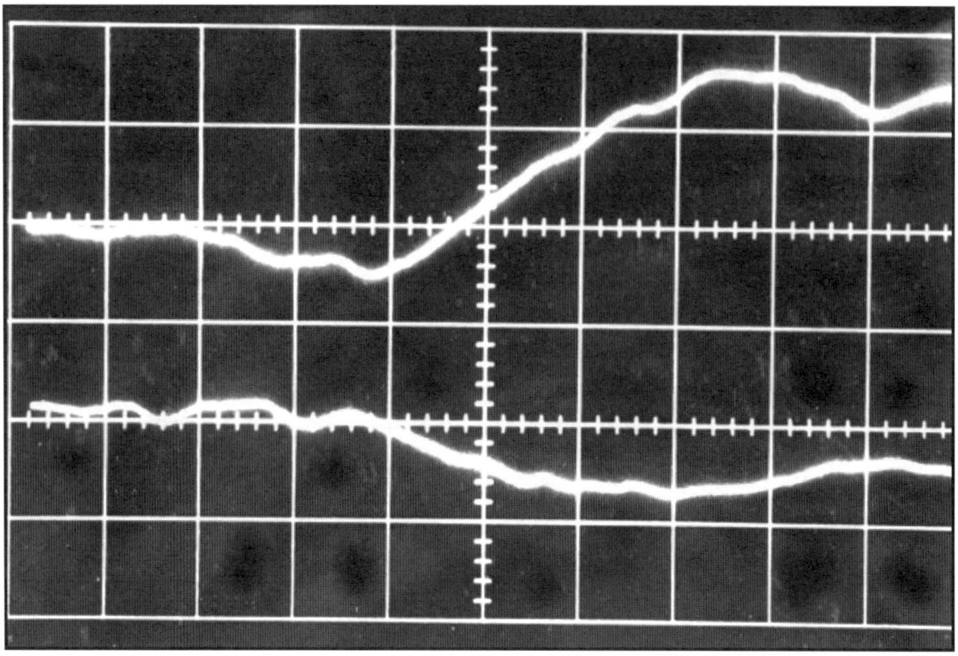

Figure 4-2 - Oscillograph record showing receiver ring moment in the vertical plane (top) and horizontal plane (bottom). Scale is 240 inch-pounds/cm (vertical) and 0.2 msec/cm (horizontal).

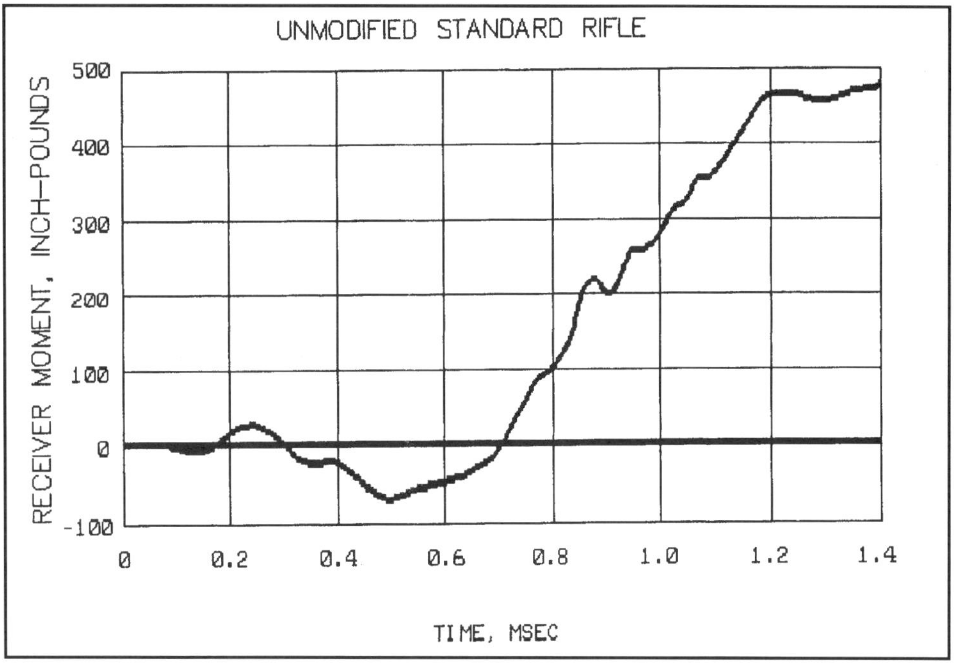

Figure 4-3 - Experimental receiver ring moment for the standard rifle with no modifications. This represents an average case.

Figure 4-2 shows a sample oscilloscope record where the cartridge is fired at 1 cm (i.e., centimeter) and the bullet exits at 8 cm on the horizontal axis. The peak moment in the vertical plane (top trace) is about 396 inch-pounds (1.65 cm trace deflection) and is about 216 inch-pounds (0.9 cm trace deflection) in the horizontal plane. This record (Figure 4-2) is representative of a case where the receiver moment is near a minimum. Normally, the lower trace is used to record an electrical signal from a switch at the muzzle that indicates that bullet exit has taken place. This bullet exit signal provides an accurate time correlation between different records. Now, I have shown the reader this sample record just so you would know what the actual data looks like. However, it is difficult to read and interpret data in this form, so from now on we will convert most of the oscilloscope traces by electronic scanning to a form that can be plotted and manipulated by the computer. This makes it much easier to add the proper scales and labels and provides a much more readable format. This results in a small loss in resolution due to computer limitations. However, the improvement in readability and understanding is well worth this small loss in resolution.

Figure 4-3 shows the receiver ring moment in the vertical plane for an average case and the plot is in the new format. Note that it shows only the moment in the vertical plane because we will concentrate on the vertical motion from now on. Note that the peak moment is about 450 inch-pounds. Based on the analysis of several hundred records, the moment can vary as much as ± 150 inch-pounds around this nominal value of 450. In other words, the peak moment can vary between 300 and 600 inch-pounds with the same load. And also, the timing of the moment pulse can vary a small amount. This variation in moment is an important effect that is caused by a number of problems that we will investigate. Later in this chapter we will use this information to estimate the effect of barrel vibration on group size.

Now, 'the name of the game' is to reduce this moment to as near zero as possible so that the barrel doesn't vibrate. If there is no moment in the forward receiver ring, there will be nothing to drive the barrel motion. The first cause of the moment that we will attack is the moment due to recoil forces.

Recoil Effects

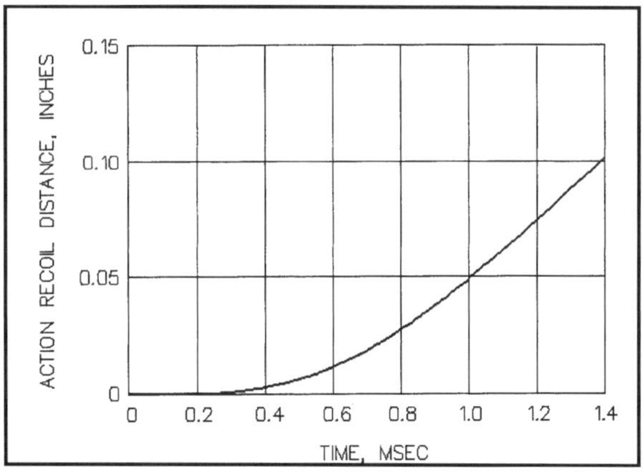

Figure 4-4 - Calculated distance that the barreled action with scope will freely recoil during the time that the bullet is in the barrel.

When the rifle is fired there is a net recoil force acting on the rifle action that is equal to the force acting on the base of the bullet, which is about 3,000 pounds at the peak chamber pressure of 53,000 psi. The force acting on the bullet was shown in Figure 2-24. This force is transmitted to the stock by the recoil lug on the bottom of the rifle action. Since there must be an equal and opposite reaction to any force, the stock exerts an equal force on the recoil lug in the opposite, or forward direction. This force results in a recoil moment being exerted on the forward receiver ring tending to drive the muzzle in an upward direction. According to computer calculations (see Figure 4-4), a rifle barrel and action, not connected to the stock, will recoil about 0.10 inch during the time the bullet is in the barrel. The recoil moment on the receiver caused by the recoil lug acting on the stock can be eliminated completely by allowing the barrel and action to recoil freely in the stock while the bullet is in the barrel. Figure 4-5 shows a picture of the Recoil Isolator which uses the principle of flexural beams that are flexible in the axial direction but are rigid in the

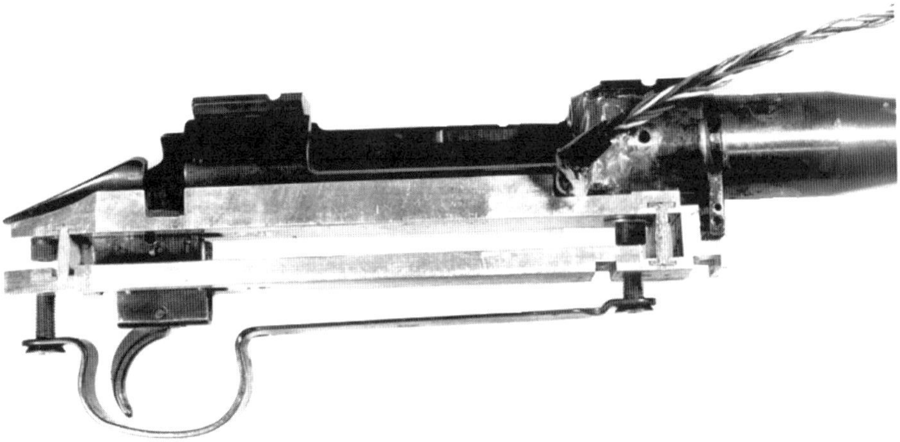

Figure 4-5 - Photograph of the instrumental rifle mounted in the Recoil Isolator.

vertical and horizontal directions. The top piece is machined to fit the receiver and is bolted to the receiver with short screws that do not interfere with the bottom piece, which is held in the stock by three short guard screws. The top piece and the receiver slide to the rear as a unit until the recoil lug engages a thin (90 mil) rubber bumper between the recoil lug and the bottom piece of the device. The purpose of the rubber bumper is to prevent peening of the soft aluminum surface by the recoil lug and to lessen the shock of the suddenly applied load to the stock. There is a small coil spring in the device between the recoil lug and the bottom piece that pushes forward on the recoil lug with a force of about 20 pounds, that keeps the recoil lug forward against the forward stop on the bottom piece. The two slots in the bottom surface of the bottom piece transmit the recoil force to the stock after the recoil lug recoils into the rubber bumper on the bottom piece. The bottom piece is bedded in aluminum filled epoxy (i.e., Devcon F) in the bottom of the stock inletting. The device is completely invisible when inserted into the stock and does not effect appearance. Consequently, the recoil lug will not experience any recoil force until it has moved rearward about 0.10 inches and the bullet has left the muzzle.

It isn't necessary for the action to move to the rear the full 0.10 inch before the recoil lug strikes the rubber bumper, because the force applied to the recoil lug has to be

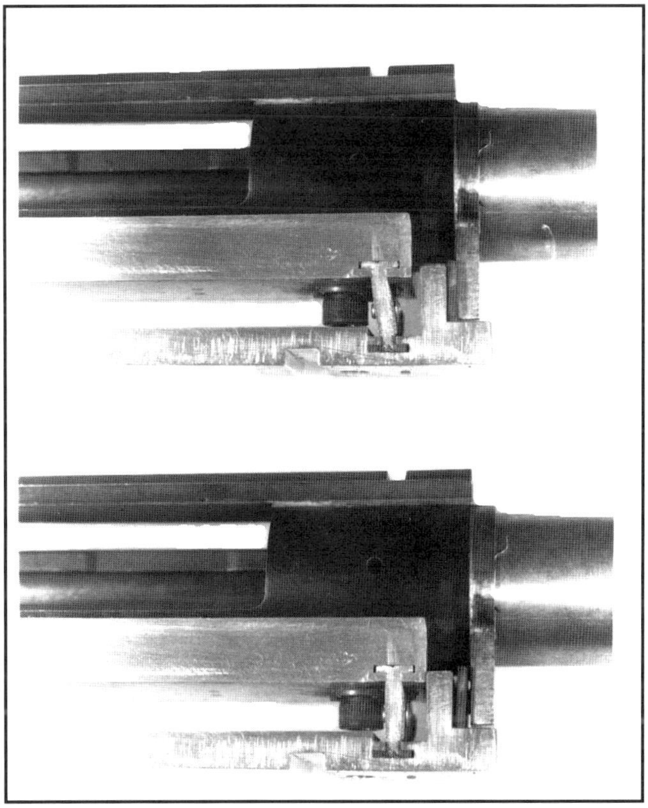

Figure 4-6 - Photograph of the front end of the Recoil Isolator showing the details of the forward flexure. The action is in the forward position (battery) in the bottom photo and in the recoil position in the top photo.

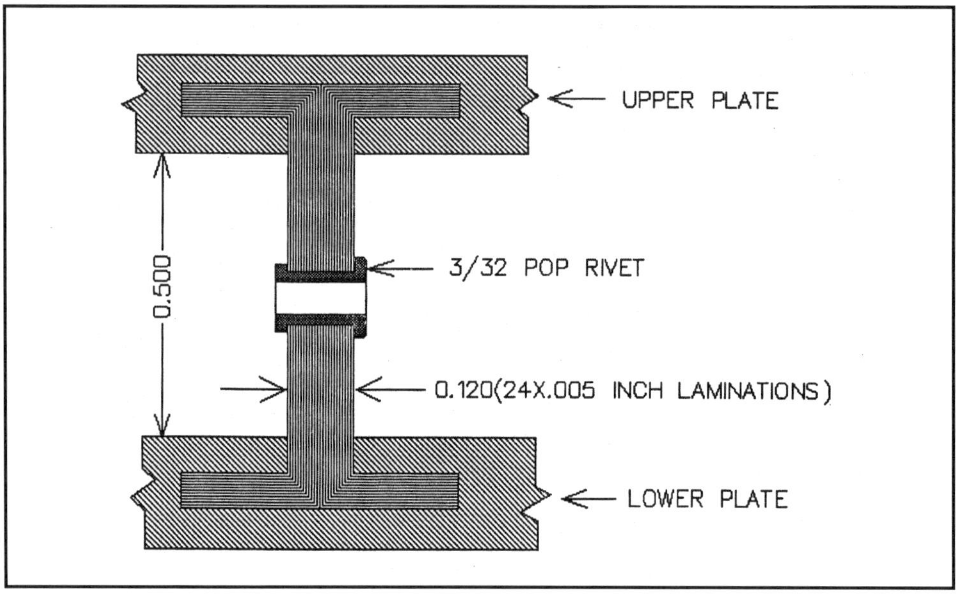

Figure 4-7 - Sketch showing a cross-section view of the Recoil Isolator laminated flexural beams.

delayed only a few tenths of a millisecond to prevent the disturbance from reaching the muzzle before the bullet exits. Therefore, one need not have more than 0.06 inches clearance between the recoil lug and the rubber bumper.

The flexural pivots are 0.120 inches thick and are made up of twenty four 0.005 inch (5 mil) thick aluminum laminations. A photograph (Figure 4-6) of the side of the Recoil Isolator shows the end of the front flexural pivot and how they bend during recoil. A cross-section drawing of one of the flexural pivots is shown in Figure 4-7. The "C" shaped laminations are made in a die on a hydraulic press and 24 of them are assembled into an "I" shaped flexural beam which is inserted into "T" slots in the upper and lower parts of the device. This laminated design is very effective in allowing a large deflection in the axial direction without exceeding the strength of the material while providing essentially the same strength and rigidity in the lateral direction that would be obtained from a solid piece of material. In this prototype the laminated flexural pivots are pinned to the upper and lower pieces to prevent horizontal motion, but at the same time allow easy disassembly. In the production case the pivots could be permanently fixed to the top and bottom pieces by staking. Also, the front of the top piece has been milled off to provide clearance for the strain gage wiring, and the machining necessary for the magazine, safety, and bolt stop were omitted in the interest of simplicity.

In production, the top and bottom pieces could be injection molded, which should make it inexpensive. This gadget is a lot stronger than it looks, and unless you are one of those people that use your rifle as a crowbar it should be strong enough. The original design did not have the 3/32 inch rivets and the flexural beams showed signs of buckling under a large compressive load. Two rivets in the front beam and one in the rear beam greatly reduced the tendency to buckle without having much effect on the stiffness in the axial direction. The height of the "I" beams was also reduced from 0.750 to 0.700 inches, because they turned out to be a little more flexible than the design calculations indicated. Reducing the height of the beams allowed adding some material to the lower plate in the vicinity of the recoil shoulder, which was marginal in strength. In spite of these minor deficiencies the original Recoil Isolator is still functioning properly after enduring the firing of more than 2000 rounds. I have gone through several designs of the Recoil Isolator and this is by far the most satisfactory.

One other design that may be worth mentioning is an application of the design used in cannon gun mounts, which is essentially a dado slide device (Figure 4-8). The problem that is inherent in this approach is that the weight of the barrel causes the rear of the top piece of the device to move up and the front to move down until the flat surfaces are engaged. This motion leaves the canted surfaces of the dado unengaged and the action can pivot in the horizontal direction. Consequently, during the 4 msec (milliseconds) that it takes for the firing pin to travel to the primer and the bullet to travel down the bore, a lot of horizontal motion can take place. This effect can be reduced by

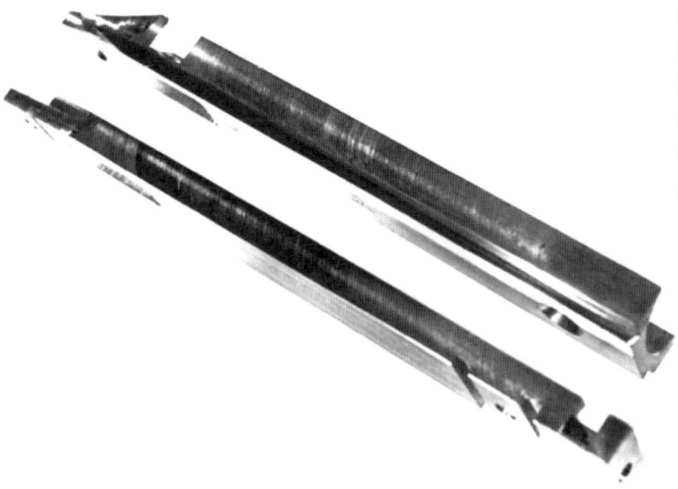

Figure 4-8 - Photograph of the Dado Slide recoil isolator device that proved to be less satisfactory than the Recoil Isolator using flexural beams.

adding compression springs between the top and bottom pieces, however, this increases the friction in the slide. In addition, there may be problems with the accumulation of dirt which could cause seizing of the slide. Anyway, I never could make this thing work satisfactorily, so it was discarded.

The forward receiver ring moment with the Recoil Isolator installed has been measured and is shown in Figure 4-9 compared with the rifle without the Recoil Isolator. You can see that the large positive muzzle up moment that we attributed to recoil has completely disappeared but it has been replaced by a sizeable negative muzzle down moment. Needless to say, I was dumbfounded by this development, because I had expected the moment to decrease to near zero when the recoil moment was eliminated. Well, it didn't, and what has happened is that we have exposed another source of moment. The problem is to find out where this remaining component of moment is coming

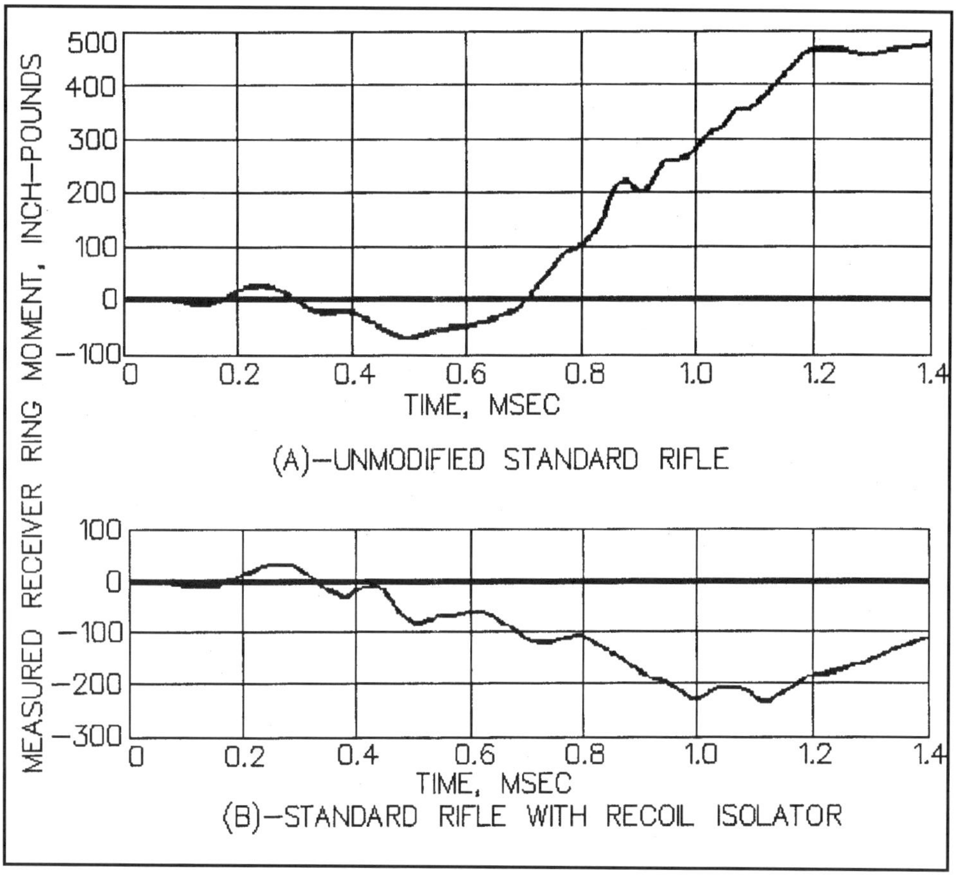

Figure 4-9 - Experimental receiver ring moment for the standard rifle (A) and the standard rifle with the Recoil Isolator (B).

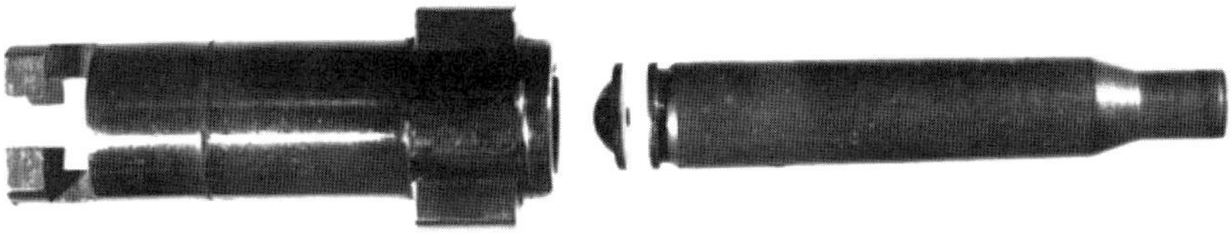

Figure 4-10 - Photograph of the rifle bolt with swiveling bolt face that proved to be unnecessary.

from. However, I will tell you in advance that we will find that the negative moment is coming from asymmetry in the forward receiver ring.

At first I thought it might be due to having a canted cartridge case head which would result in an asymmetric bolt thrust, so I built a bolt that had a swiveling bolt face (Figure 4-10) that could conform to a canted case head. The experimental data indicated that there was no effect. A theoretical analysis indicated that the brass case head is just not strong enough to transmit a large moment. This theory plus some other simpler experiments convinced me that this was not the answer.

Another idea was that the bolt thrust was not distributed equally on the two bolt lugs, so I made a bolt that had a swiveling head that had to line up perfectly with the receiver lugs (Figure 4-11). Again, there was no effect on the experimental data, so this idea was prematurely discarded. However, much later on I started working with another action that had not been fired very much, and found that the bolt lugs did not engage the receiver lugs evenly. When the barrel was removed, you could see that the bluing was nearly untouched on the top receiver lug but was completely worn off of the bottom receiver lug. Smoking the bolt lugs with a candle confirmed the fact that in a

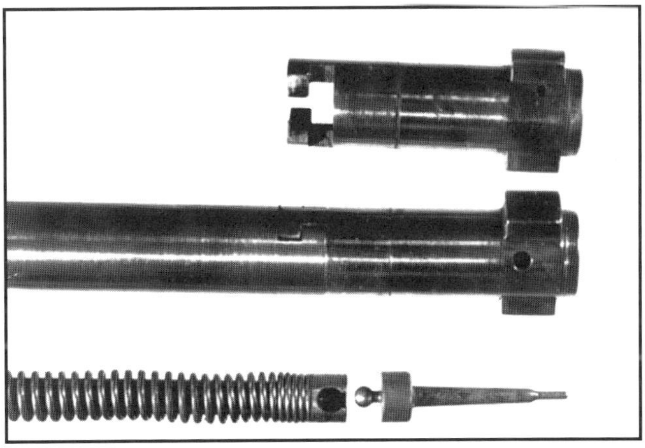

Figure 4-11 - Photograph of swiveling bolt head. Top, bolt head with double rabbet joint. Middle, assembled bolt. Bottom, swivel firing pin.

new action the bolt is canted front end down and rear end up as a result of the upward force on the sear at the rear of the bolt, which causes most of the bolt force to be absorbed by the bottom lug. Since the clearance between the receiver and the bolt body of the Remington 721 is about 8 mils the angle of the bolt cant is about 0.08 degrees. When I ran a smoke test on the old action the top and bottom lugs were evenly matched, and one wouldn't expect the swivel head bolt to make any difference in the moment acting on this old action. At this point the "light dawned". The old action—the one we are working with—had its bolt and receiver lugs accidentally lapped from firing thousands of rounds. After all, a lot of powder and primer residue, which is carbon and grit that is an effective abrasive, collects in the front of the action and promotes the lapping action. Also, I didn't clean the lugs very often. We have recently checked the bolt lug engagement on three custom bench rest actions and found that on all three actions only the bottom lug is engaged. These custom actions were tighter than regular sporters and had a bolt-receiver clearance of 5-6 mils. A lot of gun writers recommend deliberately lapping bolt lugs by applying an abrasive compound and working the action, which can be done, but it is an awful lot of work and wears some other parts of the action that you don't want to disturb. The best way to handle this problem would be for the factory to machine both the bolt lugs and the bolt face at an appropriate angle when they make the action. I modified the new Remington 721 action by machining off the rear face of the bolt lugs to 0.08 degree angle and finishing with a little lapping with the firing pin assembly in place. Lapping the lugs with the firing pin assembly removed (recommended by some gunsmiths) does no good at all, because the firing pin spring acting on the sear is what causes the rear end of the bolt to tip up in the first place. Correcting the bolt lug engagement requires removing about 1 mil from the bottom lug. At the same time I machined the 0.08 degree angle into the bolt face with an end mill. However, I recommend lapping the angle in the bolt face with an aluminum rod charged with coarse grit. The bolt is held at the proper angle in the lathe with a milling attachment. As a result the case heads stay flat after firing like they should be. If you check cases that have been fired a number of times in an unmodified action by placing a straight edge against the case head, you will find that the case heads are round. This is caused by the canted bolt face and supports the contention made earlier that the case head is too weak to cause a large moment. Just how much effect this canted bolt has on barrel vibration I cannot say, because I did not start with a new action where one could conclusively measure the effect of uneven

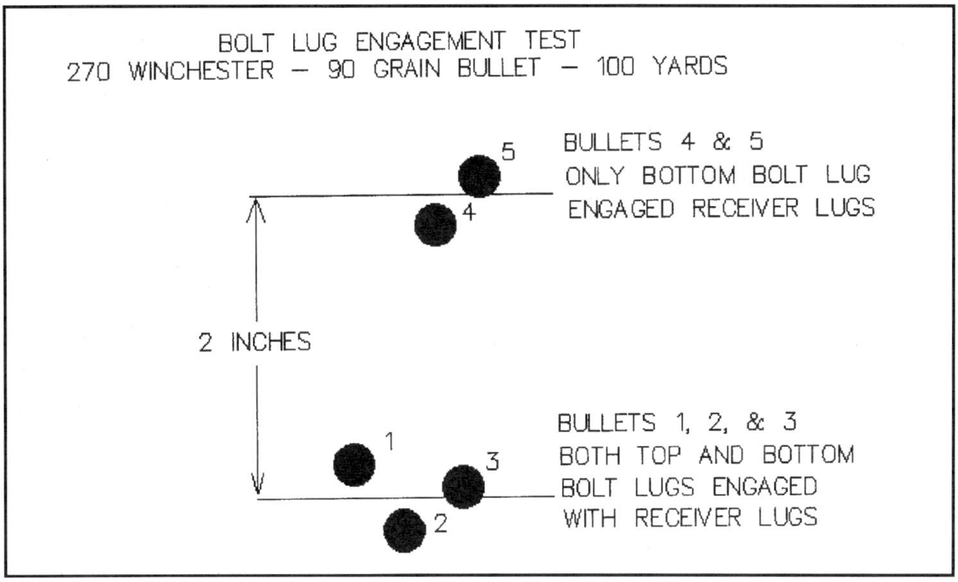

Figure 4-12 - Target at 100 yards showing how the bullet impact shifts upward when the rear surface of the top bolt lug is filed off so that only the bottom bolt lug engages the bottom receiver lug.

bolt lug engagement. However, I did run a qualitative test by firing a three shot group with even lug engagement and firing two shots after filing a few mils off the top lug. You can see in Figure 4-12 that the impact of the bullets was about two inches higher when only the bottom lug was engaged. If the bolt thrust were perfectly uniform from shot to shot, uneven bolt lug engagement would not cause a problem, but we know from strain gage measurement that this is not the situation.

As a matter of interest I measured the clearance between the bolt body and the receiver rings in an old 98 Mauser and a pre '64 Model 70 and both had a clearance of about 8 mils. It is also interesting to note that bench rest shooters are constantly cleaning the inside of their actions so that grit never has a chance to accumulate on the lugs so that "accidental" lapping doesn't have a chance to occur. Another thing that often occurs in bench rest shooting is vertical stringing of groups. The traditional medicine for this problem is to keep increasing the load (and pressure) until it stops, and sometimes it works. What is happening is the cases lengthen until they are in firm contact with the bolt face and this helps to keep the bolt thrust uniform. This is a dangerous and often unsuccessful approach to solving the problem. Some gunsmiths try to reduce the bolt clearance by sleeving the bolt. Again, you can't reduce it to zero (and that is what it would take), so that approach doesn't really solve the problem. It is much easier and more satisfactory to solve the problem by removing metal off the bottom lug until even lug engagement is obtained. It should be pointed out that vertical stringing in bench rest guns can also be caused by high frequency barrel vibration which is discussed at the end of the chapter. Now we return to the original investigation.

When I couldn't detect a problem with the bolt, I had what I thought had to be the right idea. When the cartridge case is inserted, it likely lies in the bottom of the chamber and it will have to expand radially to fill the chamber. In the process of radial expansion the case head will have to rise to the center of the chamber and through friction on the bolt face introduce a downward force on the chamber. This would give rise to a negative moment and the whole scenario seemed logical. Well, the only thing to do was to build a bolt where the bolt face is supported by several slender beams that can bend in the lateral direction but are "hell for strong" in the axial direction. I built the bolt shown in Figure 4-13, which has a slotted cylindrical insert in the bolt head. This insert acts like a wire brush—it is very strong in the axial direction parallel to the "wires" but deflects easily in the lateral or perpendicular direction allowing the cartridge case to seek its preferred position without exerting much of a lateral force. Static bench tests confirmed that it worked properly, and I proceeded to test fire it. Sure enough, the moment disappeared and I was elated. However, I became suspicious when I could not reliably repeat the test results, and I found that the two strain gages on the bottom of

Figure 4-13 - Photograph of bolt head insert that allows lateral translation of cartridge case head.

the receiver had debonded due to oil accumulation. This is a common problem with strain gages and you have to run checks continually to make sure that the gages are really working right. Unfortunately, this particular strain gage configuration is difficult to check and as a result I was getting bad data. After the gages were replaced, the negative moment returned and I had to discard this idea. All of this work took nearly two years and I was beginning to wonder if I would ever find the source or sources of the remaining moment. Then I decided to go back and look at action asymmetry as the possible cause.

Action Asymmetry

This is something that I had considered earlier but discarded because it seemed to act in the wrong direction. However, it turned out that receiver asymmetry can cause either a positive or negative moment depending on cartridge case dynamics. So, how does it happen. Figure 4-14 is a cross-section view of the receiver and barrel in the vicinity of the forward receiver ring. Most of the bending takes place in the area between the rear face of the bolt lugs and the rear face of the barrel. This is, of course where the strain gages are located.

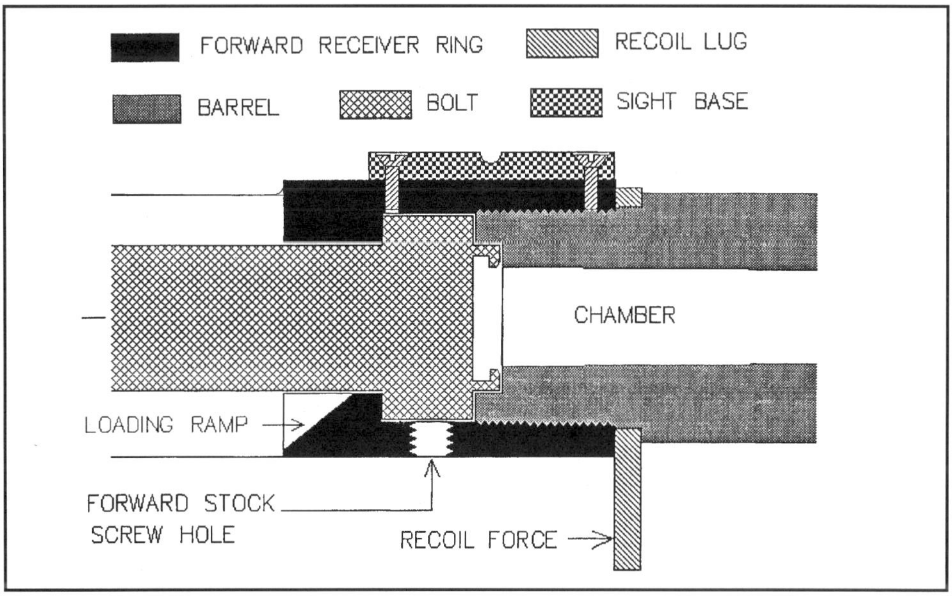

Figure 4-14 - Cross-section drawing of forward receiver ring and chamber section of barrel showing structural asymmetries in the vertical plane.

Now, notice that the receiver ring is unsymmetrical from top to bottom, and while it doesn't show in this view, it is also unsymmetrical from side to side. First, there is a 1/4 inch threaded hole for the forward stock screw in the bottom of the receiver ring, and second, there is a front scope sight base bolted on top of the receiver. Consequently, the receiver is stronger on top than on the bottom. When the case head presses to the rear on the bolt face the receiver ring stretches more on the bottom than on the top and this corresponds to a positive (muzzle up) moment. This condition is present with a neck resized case or lubricated case or when there is little or no headspace. However, when the case is short or has significant headspace, and there is no lubricant present, only the primer will contact the bolt face and a large percentage of the recoil force generated by the bullet is transmitted by a compressive force acting through the receiver ring. This results in the bottom of the receiver ring being compressed more than the top, which results in the negative moment shown in Figure 4-9(B). The fact that a short case (i.e. excessive headspace) will stick to the chamber walls and have only the primer contact the bolt face is confirmed in Figure 4-15. This figure shows the heads of three cases with headspace measurements of 15, 8, and 0 mils and you can see that the primer protrusion is proportional to the headspace. Also, the bolt thrust force was measured using the strain gages used for measuring moment by connecting them in a different arrangement. This is done by putting the two strain gages in opposing arms of the bridge. Figure 4-16 shows the condition of maximum bolt thrust that occurs with zero headspace.

Figure 4-15 - Photograph of fired 270 Winchester cases showing primer protrusion resulting from excess headspace of 15, 8, and 0 mils from left to right. Experiment is performed with special work hardened and degreased cases.

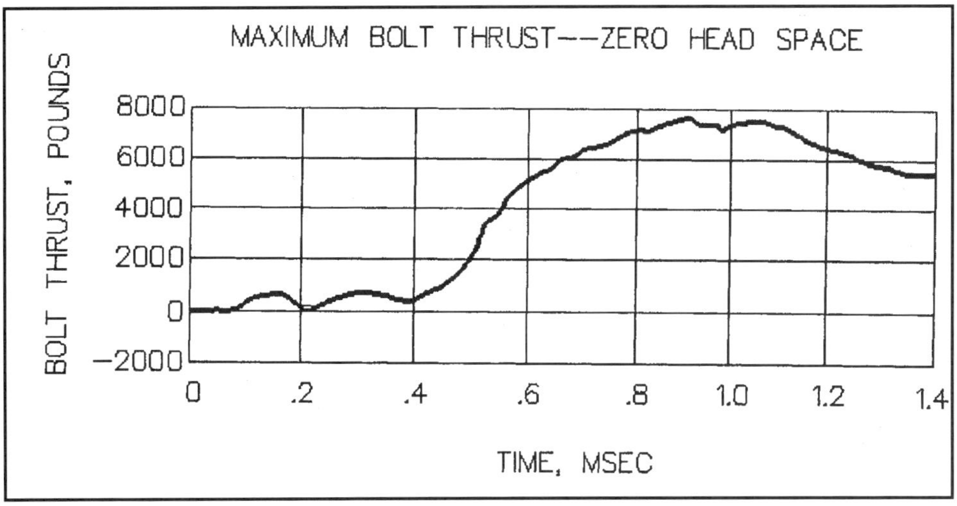

Figure 4-16 - Experimental measurement of maximum bolt thrust that occurs with either zero headspace or lubricated cases.

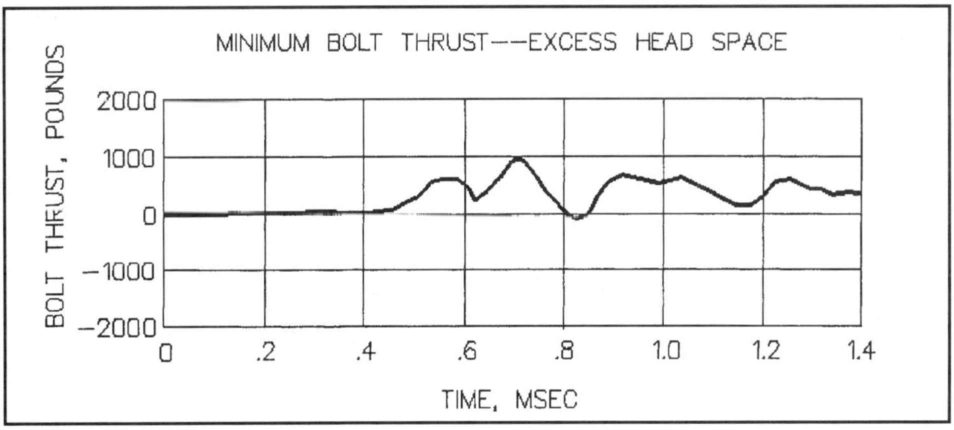

Figure 4-17 - Experimental measurement of minimum bolt thrust that occurs with headspace greater than 5 mils and degreased cases.

Figure 4-17 shows the condition of minimum bolt thrust which occurs with a large headspace (i.e., 10 mils). The measurements show that the maximum bolt thrust is about 7,500 pounds and the minimum bolt thrust is about 800 pounds. Calculations made by multiplying the internal cross-section area of the case and the primer by the peak chamber pressure show that the maximum value should be 7,200 pounds and the minimum should be 1,200 pounds. The minimum measured bolt thrust (i.e., 800 pounds) is less than the calculated value of 1,200 pounds, because the calculation does not include the effect of friction between the side walls of the primer and primer pocket. Since the recoil force is about 3,000 pounds, the receiver ring must be in compression when the recoil force is at the minimum. The data in Figure 4-17 were obtained by firing cases that were degreased and had an excess headspace of 10 mils and gripped the chamber walls. In practice you can get just about any combination of tension or compression, depending on the hardness, lubrication and length of the case. Well we can eliminate the receiver asymmetries to a great extent by making some simple modifications.

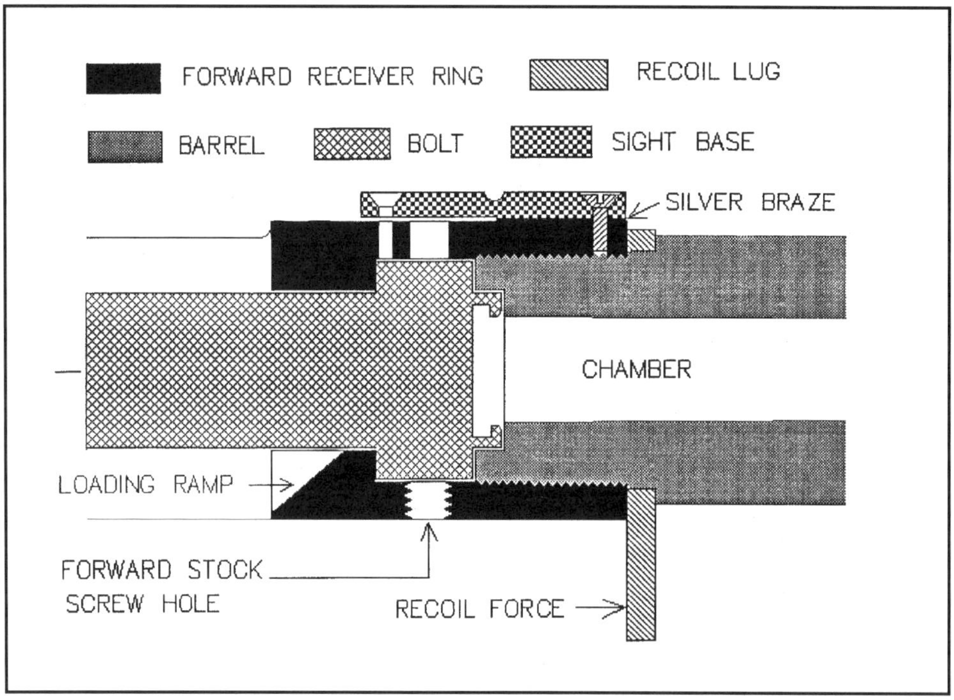

Figure 4-18 - Cross-section view of the receiver ring modified to reduce structural asymmetries in the vertical plane.

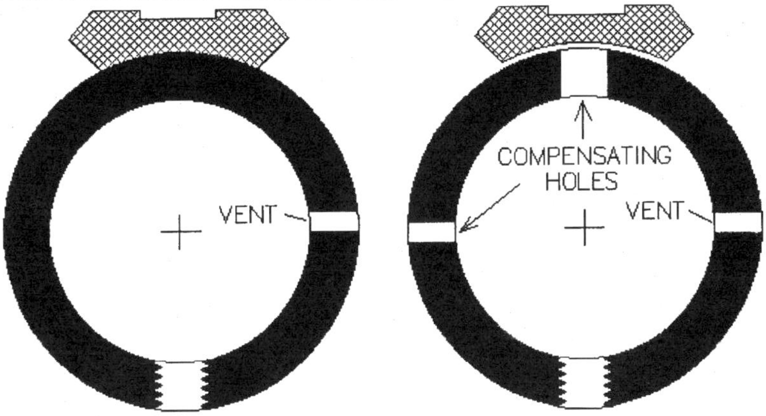

Figure 4-19 - Lateral cross-section view of the forward receiver ring and scope mount base showing modifications made to improve symmetry.

Receiver Modifications

The modified receiver is shown in Figure 4-18. You can see that a 1/4 inch hole has been drilled in the top of the receiver ring to match the guard screw hole in the bottom of the ring, and an 1/8 inch hole has been drilled in the left side to offset the vent hole on the right side of the receiver. Also, the front half of the front scope mount base is silver soldered to the receiver and the rear screw in the front base has been omitted. Now, I should tell you that the scope mount bases are made of steel instead of aluminum. This makes it possible to use a low temperature (430 °F) melting point silver braze to firmly attach the bases. Silver solder or braze is nearly as strong as mild steel and it is the only way that I have found to keep screw mounted scope sight bases from moving in the horizontal direction. Some of the newer rifles provide a direct clamping of the bases to the receiver and probably eliminate this problem. The reader should be cautioned not to try either drilling the blank holes in the receiver ring or brazing the scope sight base onto the receiver, because both these modifications are potentially dangerous as they have not been safety tested. While analysis indicates that both of these modifications are safe enough when properly performed, the only way to really test for safety is to test several modified rifles to destruction. Such tests have not been performed. Figure 4-19 shows a cross-section view of the regular forward receiver ring and the modified receiver ring, and you can see that the unmodified receiver is unsymmetrical about a horizontal axis while the modified receiver is symmetrical.

Later in the investigation I discovered that the scope sight has the effect of strengthening the top of the receiver, because it is attached to both the front and rear receiver rings. The solution would be to build a forward scope mount that would allow the scope tube to slide in the axial direction, thus preventing any application of a moment to the front receiver ring. Scope mounts for target rifles are made that allow the scope to slide in the axial direction, however they are too bulky for a sporter. I tried a simple modification to a standard Weaver mount that didn't work well, so I decided to let this problem go for a while. Since it is only a small effect we can worry about it later.

By now everyone must be wondering if all this work has had the desired effect—that is, is the receiver ring moment smaller? Well, you can see in Figure 4-20(B) that the forward receiver ring moment has been greatly reduced to a very low level compared to the unmodified standard rifle (Figure 4-20(A)). The average level of the moment acting on the modified rifle varies between +10 and -18 inch-pounds and it doesn't change much between the two extremes of bolt thrust. When you recall that we started with a peak moment of +450 inch-pounds (Figure 4-3 and Figure 4-20(A)), this represents a remarkable improvement. Consequently, barrel vibration should have been greatly reduced. However, I won't be completely satisfied until we actually measure the vibration of the barrel at the muzzle and find out for sure that we have reduced it. This can be done with an instrument called an accelerometer.

Acceleration Measurements

Measuring the acceleration on the muzzle of a rifle barrel turned out to be a very difficult problem requiring several months of effort to obtain reliable data. In fact, it involved so much work that I haven't seen before, I decided to include the technical details in Appendix A. Hopefully, if we avoid most of this technical detail at this point in the book, the reader will have a clearer picture of the results. However, we will have to discuss the instrumentation to some extent if the reader is to have a complete understanding of the data.

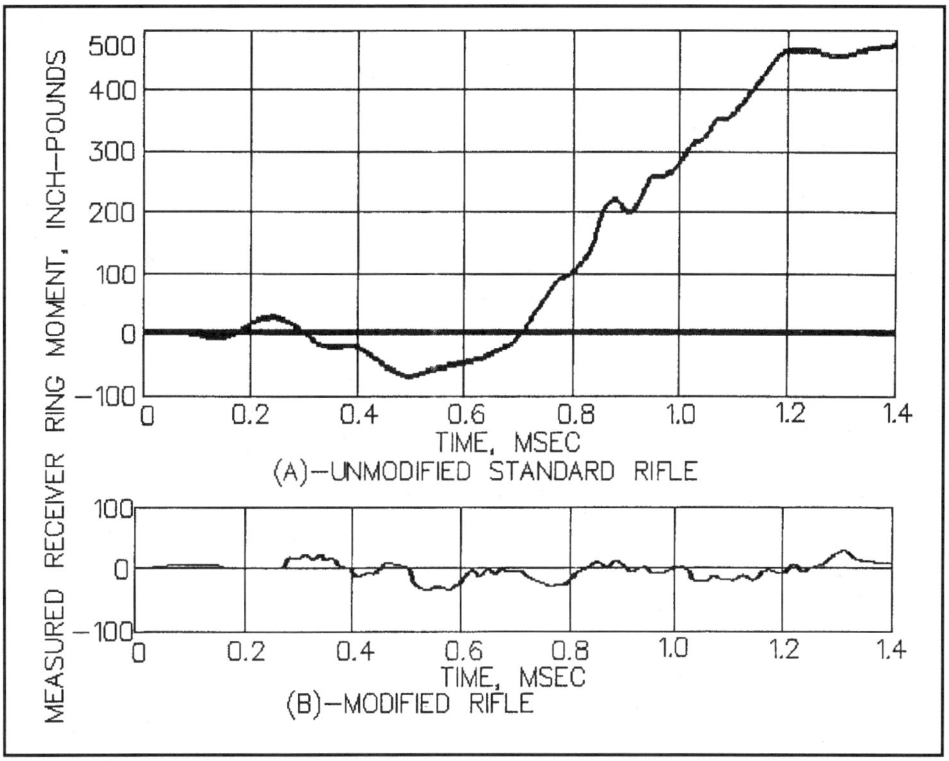

Figure 4-20 - Comparison of the measured receiver ring moment on the unmodified standard rifle (A) and the modified rifle with Recoil Isolator and modified receiver ring to achieve symmetry (B).

An accelerometer is a device that puts out an electrical voltage that is proportional to the acceleration it is subjected to. Acceleration is nothing more than the rate of change of velocity. When you speed up your car you feel acceleration. The reason we want to measure acceleration is that when acceleration is multiplied by time you have velocity, and when velocity is multiplied by time you have deflection. This process is called integration and is easily done with an electronic circuit. Velocity and deflection are the quantities that we need to tell us just how the muzzle is moving. The sensing element in this accelerometer is a thin beam mounted parallel to the bore that is 0.4 inches long, 0.2 inches wide, and 0.015 inches thick. This sensing element has a piezoelectric film bonded to both sides of the beam, which is very sensitive to being stretched or compressed, and produces an electrical signal when this little beam bends as it is subjected to acceleration. The voltage is amplified by an op-amp integrated electronics chip. The output voltage of the accelerometer amplifier, which has a gain of 50, is fed to a bandpass filter which suppresses signal frequencies both below and above a frequency of 1.25 kc (kilocycles). We will find out later that 1.25 kc is the frequency of the third mode of vibration and that it is the predominate mode of vibration. If you look back at the moment data in Figure 4-2 you can see that vibration is composed of several frequencies including the third mode, which is the lowest frequency component. There are other higher frequencies present that have little effect on the motion. The general idea is to get rid of these high frequency components in the data, because they don't contribute much to the actual motion, but they do obscure the part of the data that we want to see. The accelerometer is deliberately designed to suppress high frequencies since it has a natural frequency of 2 kc and is heavily damped with a 0.5 damping factor. The output of the bandpass filter and the integrators is recorded on the oscilloscope just as the moment data were. The biggest problem in making these accelerometer measurements is called cross-axis sensitivity. An accelerometer is never made quite perfectly and as a result it not only measures acceleration on the intended axis but picks up some of the acceleration acting perpendicular to the intended axis. This means that the accelerometer is likely to be influenced by the large axial recoil acceleration (500 G's) and indicate some small percentage of this off-axis acceleration. The bad part about this is that you can't distinguish between the cross-axis effect and the vertical muzzle acceleration data that you are trying to measure. The best commercial accelerometers have cross-axis sensitivities of 5% and according to bench tests this one has 5% to 7%. That means that if the accelerometer were rigidly attached to the muzzle, we could see as much as 25 to 35 G's from this error source. Since the vertical acceleration at the muzzle that we are trying to measure is only about 25 to 30 G's, we have a problem because the error in the measurement is as large as the value we want to measure. Fortunately, about 95% of this cross- axis effect can be eliminated by allowing the barrel to move with recoil while the accelerometer remains almost stationary, which means we will have an error of maybe 1 or 2 G's. Figure 4-21 is a

Figure 4-21 - Photograph of the accelerometer mounted on the barrel near the muzzle.

photograph of the accelerometer mounted on the barrel near the muzzle. It is a cylinder with a hole in the center that closely fits a cylinder turned on the barrel for a length of 1.5 inches. There are two spring loaded plungers located at ±45° from the bottom of the cylinder that force metal-to-metal contact. Consequently, friction between the barrel and cylinder is the only thing that can accelerate the accelerometer in the axial direction. The spring tension is reduced to the point where there is just enough force to keep the accelerometer in firm contact with the barrel when it is being vibrated at 30 G's in the lateral (i.e., perpendicular to the bore) direction. Fortunately, the barrel only has to move to the rear about 0.1 inch before the bullet exits the muzzle. A circular washer is clamped on the very end of the muzzle to prevent the accelerometer from sliding off the end of the barrel. In this way, we are able to greatly reduce the cross-axis effect. The unit of gravitational acceleration, which is labeled G, is 32.16 feet/sec^2. So now we have an accelerometer, and what does it tell us?

Figures 4-22, 4-23 and 4-24 show vertical acceleration, velocity and deflection at the muzzle for the unmodified standard rifle compared with the **modified** rifle with the Recoil Isolator and modified forward receiver ring.

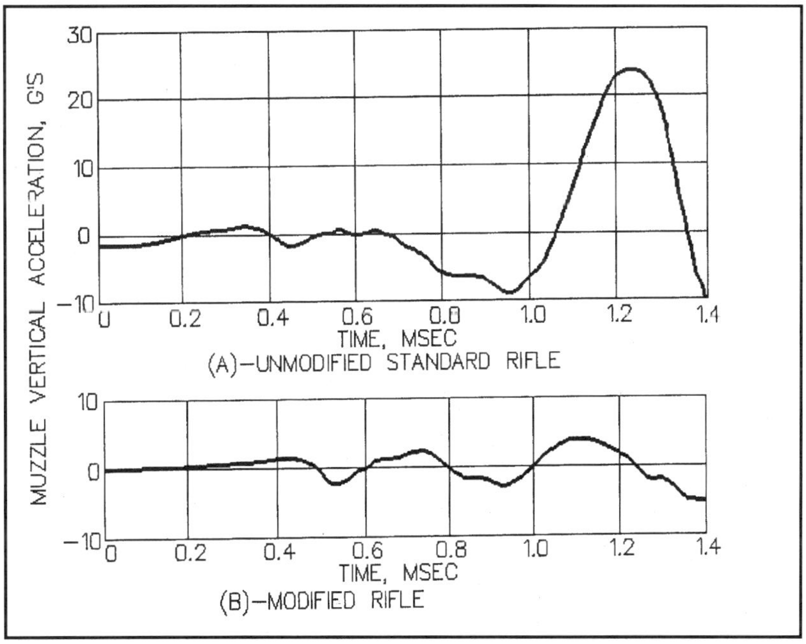

Figure 4-22 - Experimental measurements of muzzle vertical acceleration for the standard rifle (A) with no modifications and the rifle with Recoil Isolator and modified receiver ring (B).

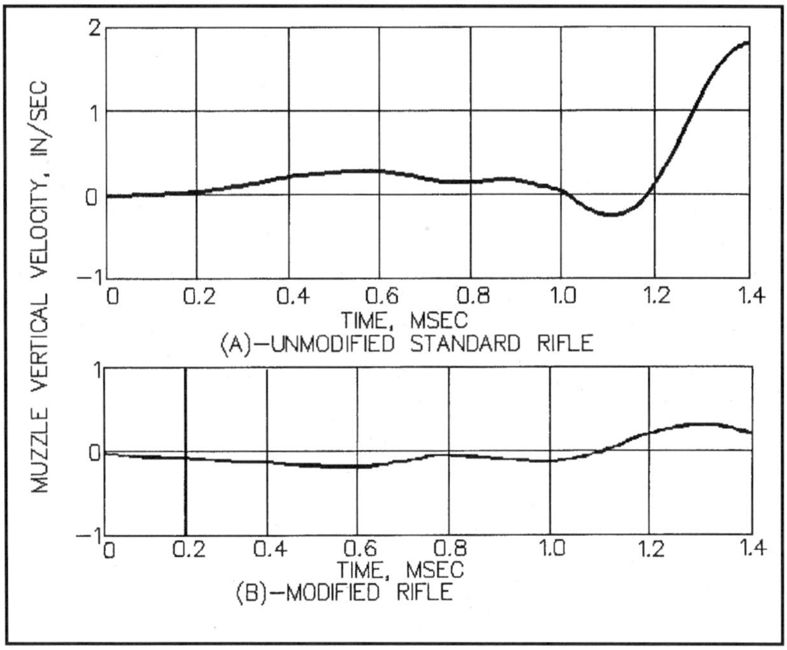

Figure 4-23 - Experimental measurements of muzzle vertical velocity for the standard rifle with no modifications (A) and the rifle with Recoil Isolator and modified receiver ring (B).

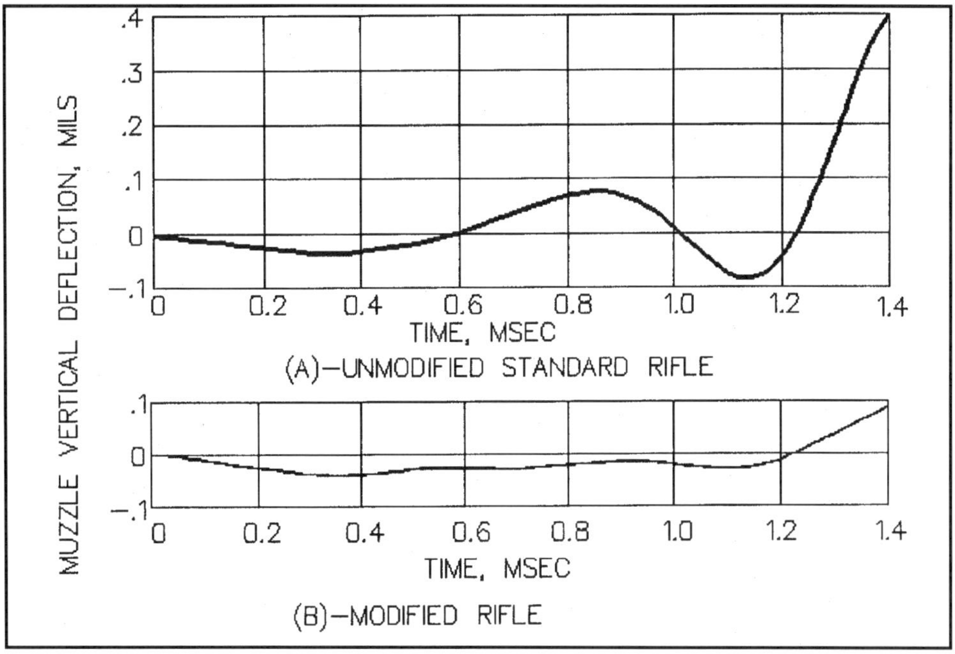

Figure 4-24 - Experimental measurements of the muzzle vertical deflection for the standard rifle with no modifications (A) and the rifle with Recoil Isolator and modified receiver ring (B).

You can see in Figure 4-22(A) that the peak acceleration on the unmodified standard rifle of about 24 G's has been reduced to about 4 G's on the modified rifle in Figure 4-22(B). Similarly it can be seen in Figure 4-23 that the muzzle vertical velocity has been reduced from 1.9 inches/sec to about 0.3 inches/sec on the modified rifle. A muzzle vertical velocity of 1.9 inches/sec may not sound like much motion, but we will see later on that it is enough to cause the bullet to shoot high by 1.4 inches at 100 yards compared to an undisturbed situation. In Figures 4-24(A) and (B) the muzzle vertical deflection for the unmodified standard rifle is compared with the modified rifle and you can see that the magnitude of the muzzle deflection has been reduced by a factor of about six. While a factor of six represents a big reduction in barrel vibration, it is considerably less than the factor of twenty or so that we saw in the moment data. Well, if you examine the acceleration (Figure 4-22(B)) data you can see that most of what we are seeing is high frequency (2.4 kc) stuff at a peak amplitude of around 4 G's and the predominate 1.25 kc third mode is not detectable. So, the predominate mode of vibration, which is the mode that we are most interested in, has been suppressed by a factor of far more than six, just as the moment data indicated. To make it easier for you to see this in the data, the periods (i.e., time between peaks) is 0.8 msec for 1.25 kc and 0.42 msec for 2.4 kc. Therefore, the accelerometer data is telling us the truth and all we are seeing in the data is the high frequency modes that were not completely filtered out. The reader is probably getting confused by this mode of oscillation discussion, so I have prepared a figure (Figure 4-25) that shows physically how a cantilever beam vibrates in the first five modes.

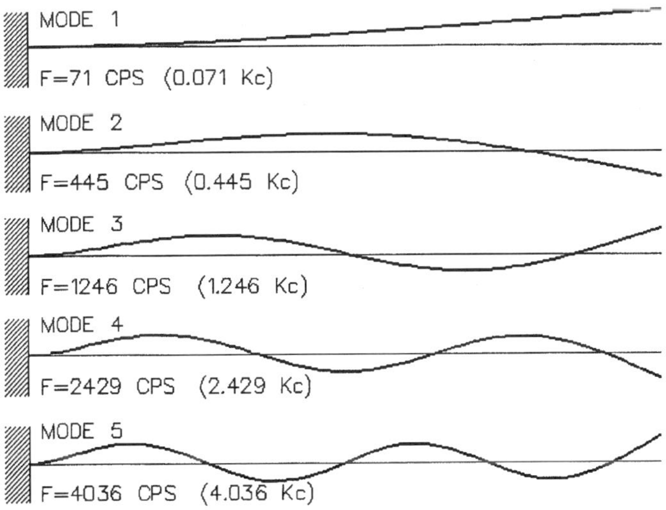

Figure 4-25 - Diagram showing how the barrel vibrates in different modes.

Also, the reader is probably wondering what all this discussion about mode of oscillation is about, and whether it has any practical bearing on the problem. Well, just have faith, it is important and not just of academic interest. What we are seeing in the accelerometer data for the unmodified standard rifle (Figure 4-22(A)) is the predominate mode 3 with traces of mode 4 and 5. In the modified rifle acceleration data (Figure 4- 22(B)) we are seeing primarily mode 4 with a trace of mode 5, which means that mode 3 has been effectively eliminated, which has been our objective. The reason for this is that we eliminated the forces and moments that are capable of driving the first through the third modes, which were relatively slowly varying forces, but did not eliminate the forces that drive the higher numbered modes. You see, you can't force a beam to oscillate at frequencies significantly higher than the frequency of the driving force. Since the driving moment that we eliminated is directly related to the chamber pressure, which has a fundamental frequency roughly equivalent to mode 3, we should expect the first three modes to be excited, but not the higher modes. That means something else is disturbing the barrel at the higher modes. Well, the way to find out what that something else is, is to operate the rifle on the bench without firing a cartridge and make the same measurement. Figure 4-26(B) shows the

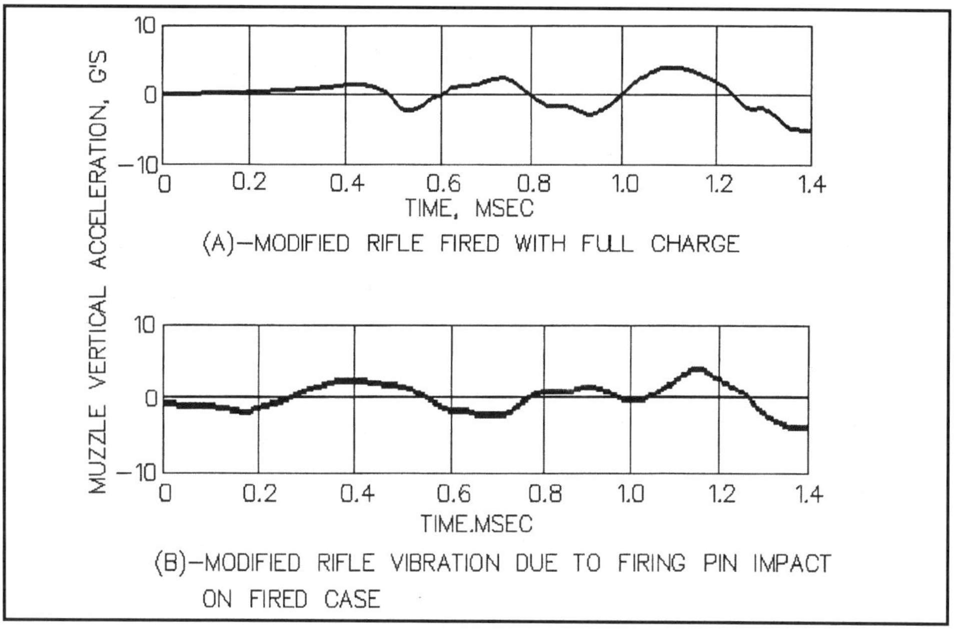

Figure 4-26 - Comparison of the experimental measurement of muzzle acceleration on the modified rifle fired with a live cartridge (A) and the muzzle acceleration caused by the firing pin impact on an empty cartridge case (B).

acceleration data obtained by simply pulling the trigger with a fired case in the chamber of the modified rifle. It looks remarkably like the acceleration data on the modified standard rifle fired with a loaded round in Figure 4-26(A). This means that the firing pin impact on the primer is driving these high frequency modes. An impact force has a very high frequency content and is capable of driving the barrel at high frequencies. So, how do we deal with the firing pin impact problem?

Firing Pin Impact

The only effective way that I know to eliminate firing pin impact is to hang your rifle on the wall and don't shoot it or use electric firing. While I have fired primers electrically, there doesn't appear to be any feasible way to apply it to a field rifle. However, 'not to worry', because you can prove that firing pin impact is not a significant contributor to dispersion. The maximum vertical velocity that can result from a 4 G acceleration at a frequency of 4.2 kc acting over one half cycle (time = 1/4200=0.00024 sec) is

muzzle vertical velocity = 4x32x12x0.00024 = 0.37 inches/sec.

where the muzzle vertical velocity is obtained by multiplying the acceleration in inches/second by the time for a half cycle. Note that on Figure 4-23(B) the maximum possible value of the vertical velocity is about 0.3 inches/sec on the modified rifle, which agrees with our calculation. The maximum deflection of the bullet at 100 yards due to a muzzle vertical velocity of 0.37 in/sec at an average flight velocity of 3000 ft/sec (time of flight=0.1 sec) is

bullet deflection = 0.37x0.1 = 0.037 inches.

The dispersion depends on the variation in the firing pin impact force from shot-to-shot and according to what I have measured on the bench it may amount to as much as 20% to 30%. Therefore, the dispersion resulting from variations of firing pin impact force is probably no more than 0.007 to 0.010 inches (7 to 10 mils at 100 yards). As engineers say, "this is down in the mud", and not worth worrying about. Some shooters reduce the length of the firing pin spring, which reduces the firing pin impact force. Unfortunately, reducing firing pin spring stiffness also reduces reliability and increases the time of fall (e.g., lock time), which is already 2.4 msec on this rifle. Increasing the firing pin time of fall gives the rifle more time to move between the time the trigger is pulled and the bullet exits the muzzle, consequently, accuracy can only suffer. Other shooters try to reduce the lock time by increasing the stiffness of the spring, but this will increase barrel vibration and can make the bolt difficult to operate. Brownells sells a Tubb titanium firing pin that is half the weight and a spring that has a 16% increase in stiffness. This combination decreases the lock time by 35% while maintaining the impact energy roughly equivalent to the standard firing pin assembly. So, with this set up, the firing pin disturbance is the same, but the lock time is reduced. This modification may help accuracy particularly when firing from standing or sitting positions, so it may be an improvement on target rifles. Since I don't know of any simple way to evaluate the effect of lock time on accuracy, I am inclined to leave the firing pin design alone, because there is no clear evidence that making a change will reduce dispersion.

Cartridge Case Wall Thickness Asymmetry

If the cartridge case wall thickness varies around the circumference of the case, there will be a thick side and a thin side. When the case is pressurized the thin side should stretch more than the thick side in the axial direction. This should shift the bolt thrust slightly off the center line of the action, and cause a moment. The moment will, of course, make the barrel move and produce dispersion. There has been quite a bit of discussion of this problem in bench rest publications (i.e., "Precision Shooting"). I modified a Forster Coax Gage, as suggested by Olsen in the March 1993 issue of "Precision Shooting", to measure case wall thickness. After running a bunch of Remington cases through this gadget the results were as follows.

Percent of Cases	Wall Thickness Difference
35	< 0.001
15	0.001 to 0.0015
30	0.0015 to 0.0025
20	0.0025 to 0.004

The results show that about half the cases are pretty good, but about half show wall thickness differences of 2-4 mils. Now, 4 mils is about a 16% difference in side-to-side thickness, and that could make a significant difference in bullet impact. I made a rough theoretical calculation that indicated one might expect as much as 0.25 inches dispersion from this source. The only way that I know to test this effect is to take a bunch of the worst cases, and index five rounds with the thin side up and five rounds with the thin side down. If there is an effect, there should be two groups displaced in the vertical direction. Well, I tried this and the results were inconclusive. In retrospect, I decided that one might not expect to see an effect where one has a normal headspace (0.002 inches) combined with a chamber that has moderately rough walls and a spring loaded ejector that pushes the case all the way forward. You might see this effect in a bench rest gun where the chambers are highly polished and there is usually no spring loaded ejector combined with minimum headspace. In my experiment the case probably stuck to the sides of the chamber so that no effect was seen.

We now turn to the computer to determine how much dispersion we can get from barrel vibration.

Barrel Vibration Computer Simulation

A computer code was developed that accurately predicts the vibratory motion of a rifle barrel. Since the average reader will not be interested in a lot of detail on this code, only a brief description will be given. If more detail is needed, the reader can turn to Appendix B where a somewhat detailed discussion is given. The computer code is not reproduced because it is very user unfriendly. The barrel is divided into 24 elements, where the motion of each element can be described by an equation. The equations contain influence coefficients which calculate the influence of all the other elements on a single element. In this way the forces and accelerations acting on an individual element can be calculated and also the motion of the individual elements can be calculated. The code was checked by comparing the acceleration data obtained on a vibrating cylindrical beam, which verified the code.

Figure 4-27 shows a sketch of the barrel that indicates the individual elements and the circular symbols show how the barrel droops downward as a result of gravity. It may surprise you to find out that the barrel droops about 5.3 mils as a result of the action of gravity. Figure 4-28 shows the barrel deflection for progressive time steps with the gravity droop removed for clarity, and you can see that a wave forms that propagates toward the muzzle.

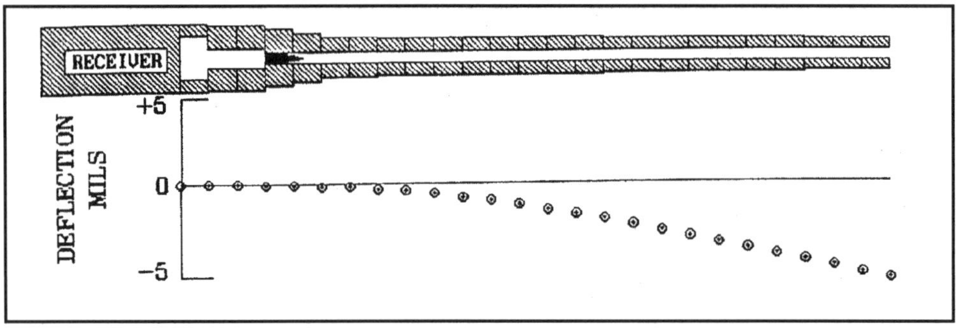

Figure 4-27 - Computer drawing showing how the barrel is divided into elements for calculation of barrel vibration by the barrel vibration computer code. It also shows how gravity causes the barrel to droop.

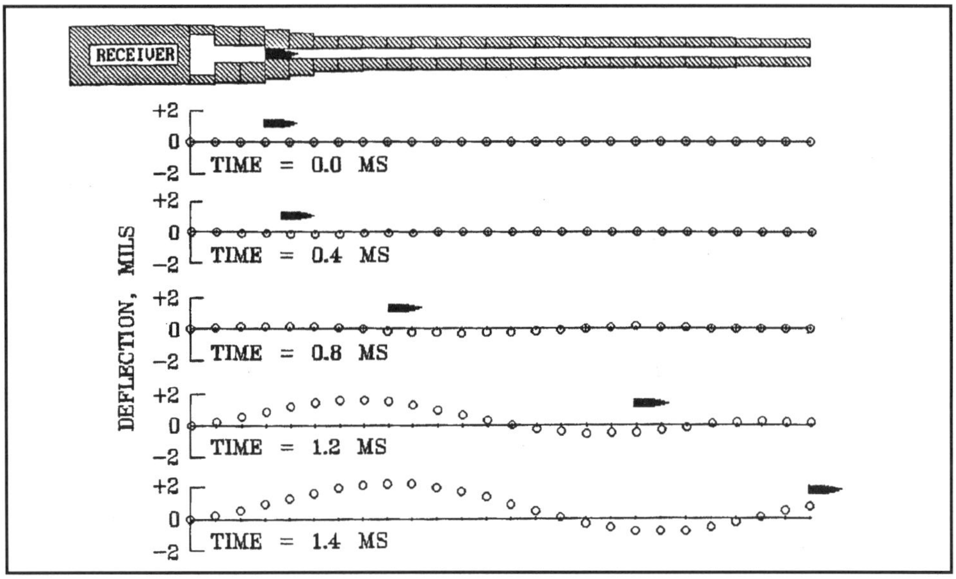

Figure 4-28 - Computer drawing showing how the barrel vibrates as predicted by the barrel vibration computer code for various selected times. The gravity droop has been removed for clarity. The bullet symbol indicates the position of the bullet in the barrel.

The scale indicates a deflection of 2.5 mils and the scale is greatly exaggerated for clarity, so you can see that the barrel does not move very much. However, it is enough to cause considerable dispersion. Note that if you compare the wave shape of the distorted barrel it agrees very well with the third mode of oscillation shown in Figure 4-25. The miss distance, which is the difference in point of bullet impact between this case with vibration and the point of impact of the undisturbed barrel, is 1.406 inches. The barrel vibration computer simulation tells us that the muzzle end of the barrel is pointed upward with an angle of 0.0175 degrees and the muzzle is moving upward at a velocity of 3.08 inches/sec when the bullet exits. The vertical velocity accounts for about 0.3 inches of miss distance and the muzzle angle accounts for about 1.1 inches of the total miss distance. Now, if the moment, and consequently the muzzle motion was the same with every shot, this change in impact point would not cause any dispersion. But, the moment can vary by about ±30% from shot to shot and this means we get about 60% of this change in impact point, or about 0.8 inches of dispersion. The variation in moment is primarily caused by variations in bolt thrust plus the other factors that have been discussed. This estimate of the variation of moment, and in effect the amount of vibration, is based on the analysis of roughly 800 oscilloscope records taken on the unmodified standard rifle. Obviously, that is a lot of data, so the estimate of the variation should be reasonably good. But, just how good is this computer simulation of barrel motion? Well, it's pretty good, as you can see in Figures 4-29, 4-30, 4-31, and 4-32, which compare the nominal experimental and theoretical values for receiver moment, muzzle acceleration, muzzle velocity, and muzzle deflection for the unmodified standard rifle, all in the vertical plane.

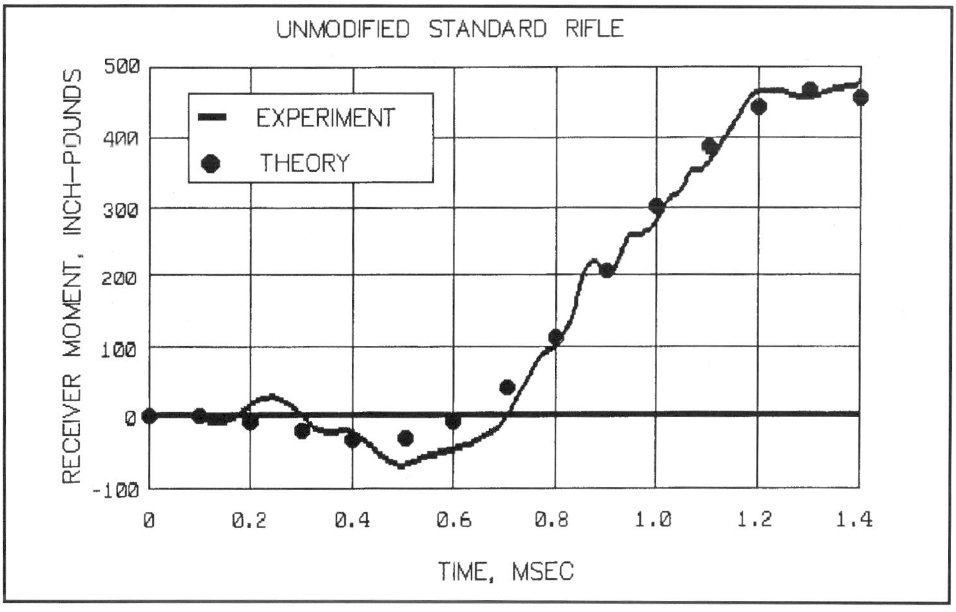

Figure 4-29 - Comparison of the measured experimental receiver ring moment for the unmodified standard rifle in the vertical plane with the calculated moment from the barrel vibration computer codes.

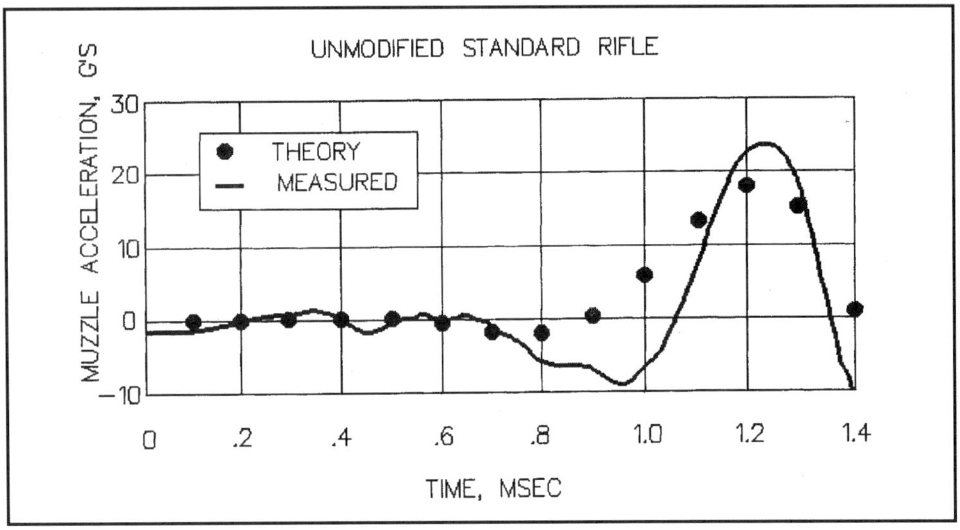

Figure 4-30 - Comparison of the measured experimental muzzle vertical acceleration for the unmodified standard rifle with the muzzle acceleration obtained from the barrel vibration computer code.

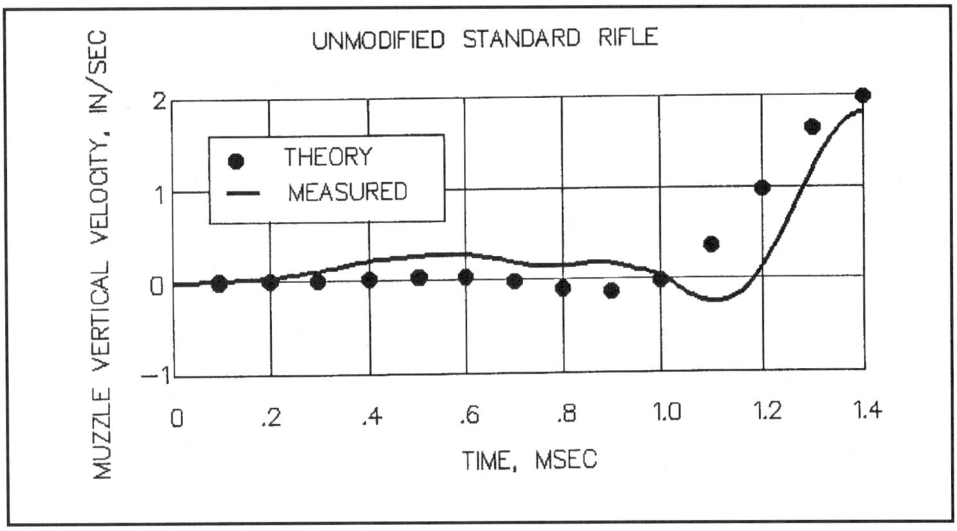

Figure 4-31 - Comparison of the measured experimental muzzle vertical velocity for the unmodified standard rifle to the calculated muzzle vertical velocity from the barrel vibration computer code.

Before leaving the theoretical work, we need to point out that the strain gages actually measure what is known as the response moment, which results from an actual applied moment. You see, the moment we measured is much smaller than the real moment that was applied by the recoil lug and asymmetric

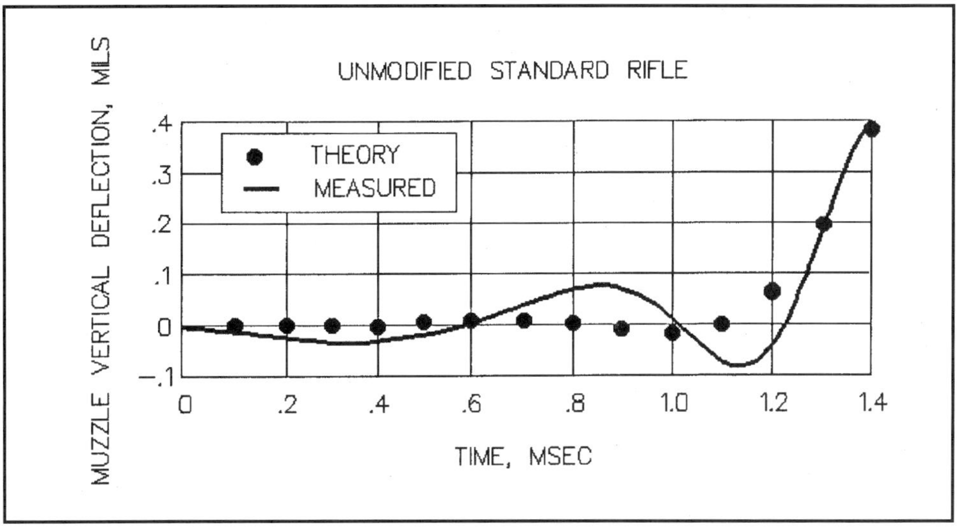

Figure 4-32 - Comparison of the measured experimental muzzle vertical deflection for the unmodified standard rifle to the calculated muzzle vertical deflection from the barrel vibration computer code.

receiver. The reason for this is that the barrel just can't respond quickly enough, so that the response moment would equal the applied moment. Both the applied moment, which peaks at about 1500 inch-pounds, and the response or measured moment, which peaks at about 450 inch-pounds, are shown in Figure 4-33. The reason for discussing this is to point out that the true moment (applied moment) that is actually forcing the barrel to move is very large, about three times the measured or response moment. Also I need to

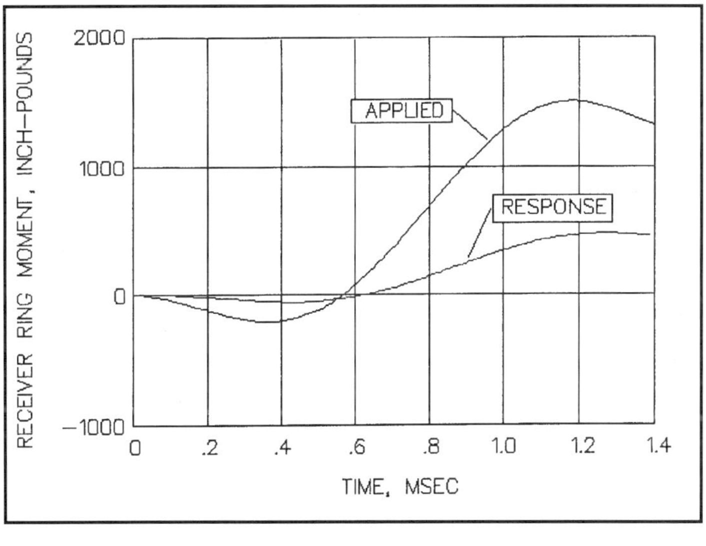

Figure 4-33 - Calculated applied and response receiver ring moment for the unmodified rifle showing how the actual applied moment is much larger than the response moment. The response moment is the average moment measured by the strain gages.

point out that the stiffening effect of internal pressure was found to be negligible. A barrel is a thick walled cylinder compared to a fire hose for example and the effect is insignificant (Reference 17). In addition, the fact that barrel gravity droop causes the bullet to travel a curved trajectory in the bore that generates an upward centrifugal force on the barrel was evaluated and found negligible.

Horizontal Dispersion

We now need to consider the effects of muzzle motion in the horizontal plane. While I did not make nearly as many measurements in the horizontal plane as I did in the vertical plane, I made enough to convince me that the horizontal motion of the muzzle is about 1/3 of that in the vertical plane (0.8 inch). Thus, the unmodified standard rifle dispersion in the horizontal plane should be about 0.27 inches. This would make sense, because the only obvious asymmetry in the horizontal plane is the vent hole (1/8 inch diameter) drilled in the right side of the forward receiver ring. This hole is supposed to vent gas escaping through a hole in the bolt head. I doubt that this vent hole really works, because the hole in the bolt head opens into the bolt lug raceway, and most of the escaping gas would flow through the raceway and out the loading port. In a new action design I would leave it out. Anyway, this asymmetry was eliminated by drilling a matching hole on the opposite side of the ring.

The reader should be warned that drilling holes in a rifle forward receiver ring may be dangerous. While a stress analysis indicates that the receiver ring still has a very large factor of safety with the two additional holes, you simply cannot be sure that the procedure is safe without testing to destruction. Such testing has not been done.

I believe that we have evaluated the barrel vibration problem with a high degree of certainty, and I also believe that we have eliminated the significant causes of the vibration. The next step is to test fire the modified rifle for group size and see if there is any improvement.

Accuracy Test

The results of the accuracy firing tests on the modified rifle with the new barrel are shown in Table 4. The tests were performed using a standard bench rest firing setup, and the results are a summary of twenty 5 shot groups. The group sizes are measured with a dial gage micrometer, and represent the extreme or largest spread between centers of the bullet holes.

TABLE 4

Accuracy Test (5 shot groups at 100 yards)

Average	Maximum	Minimum
0.884	1.223	0.408

Now, you may have expected better results, but remember that this rifle typically would shoot 1.5 inch average groups before we started the modifications. Consequently, we have improved the accuracy by nearly a factor of two, and that is real progress. Also, there are several more known errors in the rifle that can account for the remaining dispersion.

Accuracy Testing

We need to talk about the statistics involved in testing. Most people think that if you have a ballistic system (i.e., rifle) that has two error sources and you eliminate one of the errors, the resulting dispersion will be reduced by the amount of the eliminated error. Unfortunately, it doesn't work that way, and depending on the number of error sources in the system, the resulting dispersion will usually be reduced by a much smaller amount. The reason for this is that the total dispersion of a system is equal to the square root of the sum of the squares of the individual error sources.

$$\text{Total Error} = \sqrt{(A^2 + B^2 + C^2 + D^2 + \cdots)}$$

where A, B, C, and D are the individual error sources.

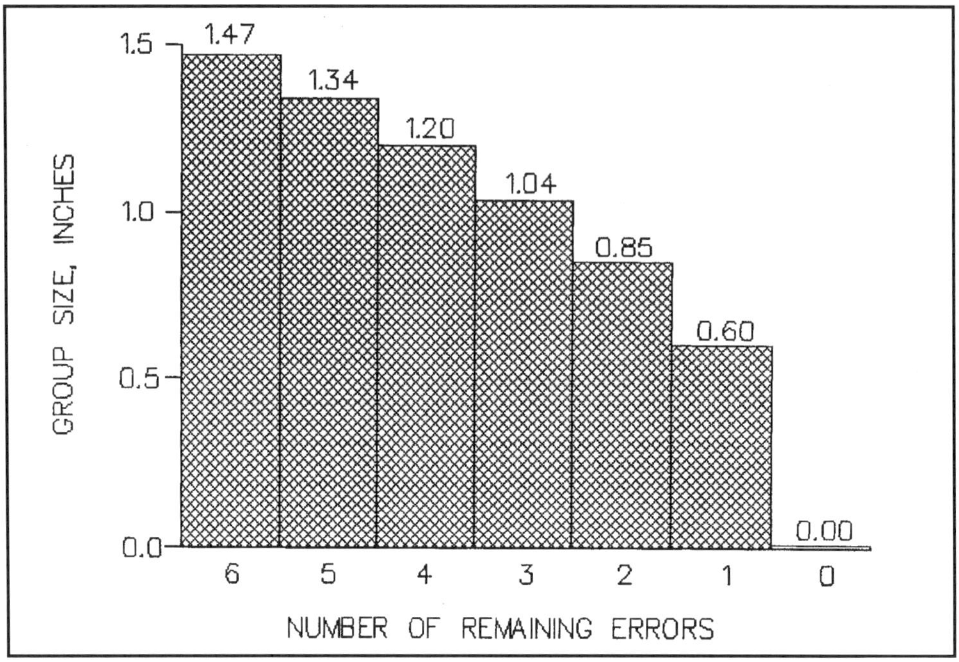

Figure 4-34 - A plot of dispersion as a function of the number of error sources remaining in the rifle where there are initially six equal errors of 0.6 inches. This demonstrates the difficulty in determining when an error source has been corrected by test firing.

Figure 4-34 shows how the calculated dispersion for a rifle with six equal sources of error (i.e., 0.6 inches) changes as you gradually remove each source of error. Note that when the first 0.6 inch error is removed, the dispersion improves from 1.47 inches to 1.34 inches and not from 1.47 to 0.87 inches as many people might expect. In other words, as long as any other significant sources of error remain in a rifle you usually can't detect the full effect of eliminating a single error. One of the best examples of this effect that I have seen happened at least twenty years ago. A very reputable ballistics laboratory was given a contract to determine the effect of bullet tip mutilation on group size. The firing tests were run using a Mann barrel, which is a large diameter (about 3 inches) barrel mounted on a concrete pylon. This configuration should eliminate sighting and barrel vibration errors, however, at least two other significant errors remained, which we will investigate in later chapters. Well it turned out that the average group sizes were around 0.6 inches, and no effect was detected. It turns out that mutilation of bullet tips does have a small effect that can only be calculated (Chapter 10), but this effect was obscured by the other remaining error sources. All that this test established was that the error was significantly less than the group size, which is something that professional ballistics people already knew. The point of all this is not to expect the group size to be improved by the full amount of error that was attributed to barrel vibration (0.84 inches), because there are still some other significant errors present. The other point to be made is that it may be difficult to tell whether or not you have eliminated a source of inaccuracy by test firing.

When I started testing this rifle I had expected to get approximately a 0.65 inch average group. When the average group sizes turned out to be significantly larger (i.e. 0.94), I decided to check the throat by making a sulphur cast of the throat. This barrel had been fired between two and three thousand times, and I was suspicious. Sure enough, the cast showed that the rifling in the throat had eroded forward by about 0.43 inches. That is considerably more than I would have expected, and too much to be corrected by setting back the barrel. Consequently, the only thing to do is to start over with a new barrel. The new barrel was made from a new Douglas blank and was chambered with the same tools used on the original, so it should have been as near identical as two barrels could be. This barrel was used in obtaining the results shown in Table 4.

Action Bedding

There has been a lot written about epoxy bedding and most of it consists of unsubstantiated claims. Contrary to all the grandiose claims, I can't see a big difference in sporter accuracy between a good inletting job and epoxy bedding. However, it may make a small difference in the case of a sloppy factory bedding job. Since it won't hurt anything, and may make you think you have done something good, you might as well epoxy bed the action if you feel like it. I have found a thin coat of plain old Elmers epoxy to be as good as anything. I don't like glass filled epoxy because it wrecks sharp chisels and I don't think it is any better. The strongest epoxy for bedding is aluminum filled Devcon F which has a putty consistency and is easier on chisels. However, it may show as a bright line around the edges of the inletting. This stuff is noticeably stronger than ordinary epoxy, if that is important. Epoxy bedding will protect the wood from deterioration from oil soaking and should cause the action to come closer to assuming the same position in the stock after each shot. But like I say, I can't really tell any difference.

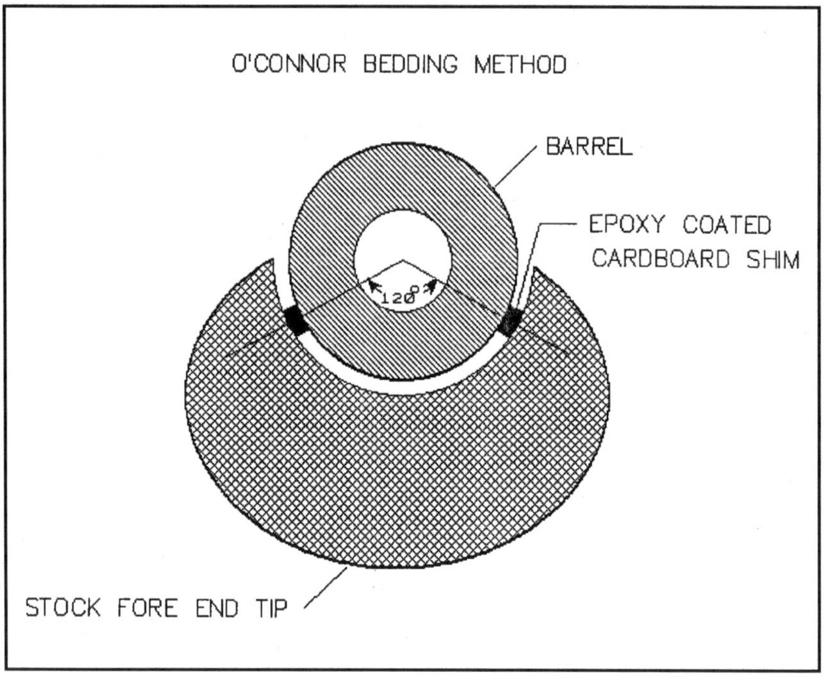

Figure 4-35 - Cross-section view of the stock forearm tip showing how cardboard shims coated with epoxy are placed in the barrel channel in O'Connor bedding approach. A differential force of 10-20 pounds is required. Barrel vibration is reduced by a factor of two over a free floating barrel.

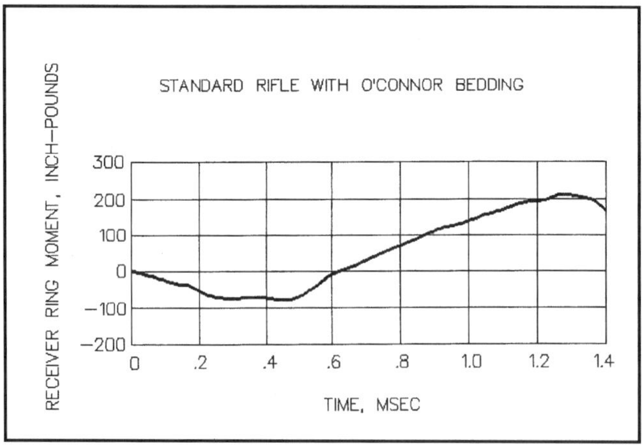

Figure 4-36 - Measured forward receiver ring moment with O'Connor bedding showing how the moment is reduced by about 50% compared to a standard rifle. See Figure 4-3 for comparison.

The only conventional bedding method that I have found that definitely improves accuracy is what I call the O'Connor method. I call it the O'Connor method because I think I first read about it in O'Connor's column some 50 years ago. Whether it was his idea or the idea of a gunsmith of his acquaintance, I can't say. But it works well enough to improve the accuracy of most commercial rifles by about 20 to 30 percent. The idea is to bed the action so that the forearm tip of the stock contacts the barrel with an upward force of 10 to 20 pounds. The barrel should be free of the stock from the action to the tip of the forearm. It may be necessary to remove some of the wood from under the front of the receiver to tip the barrel down enough to achieve the 10-20 pound load. It is best to place two epoxy coated inserts (i.e., see Figure 4-35) in the barrel channel spaced circumferentially by about 120 degrees and the receiver should be epoxy bedded. This method helps because it applies a preload moment, which is a method commonly used by design engineers. Figure 4-36 shows the forward receiver ring moment measured on the standard sporter rifle with O'Connor style bedding. If you compare Figure 4-36 with Figure 4-3, you can see that the recoil moment has been reduced by nearly 50%, which is a significant amount. Notice also that the amplitude of the high frequency oscillations has been reduced, probably as a result of the friction damping between the forearm and the barrel. The point of impact may drift downward slowly for several months after the action is bedded in this manner but it will eventually stabilize. A violent change in weather conditions, particularly humidity, could cause a shift in point of impact. However, I have carried hunting rifles from the Mexican border to the northern Yukon Territory for years that were bedded in this manner and never had any trouble. I think most of the trouble with shifting point of impact comes from using improperly seasoned and aged stock wood. A stock blank should be aged for at least five years before it is used so that it has time to stress relieve. There is a commercial device (AccuMajic Accurizer) made by Aftermarket Innovations (1-800-528-6900) that seems to accomplish the same thing as O'Connor bedding. I have not tested this device but according to an article in the February 1996 Shooter's News (1-216-979-5258) it seems to work the same way. Anyway, we now have experimental data showing why O'Connor's method works and I can recommend it for hunting rifles.

Pillar bedding has been used in bench rest rifles with free floating barrels. In this type of bedding two 1/2 to 5/8 inch diameter aluminum rods with holes for the guard screws are epoxy bonded in the stock. The upper surface of the pillars are machined to fit the bottom surface of the receiver. This type of bedding holds the receiver in the stock very rigidly, and apparently reduces flexing of the action. I tried this about 30 years ago and found that it does help with a free floating barrel. I don't like it on a hunting rifle because it makes the action noisy. These days most bench rest rifles have the actions glued in a plastic stock which is very successful.

Barrel Weight

Everybody seems to think that increasing barrel weight improves group size. The question is just how much? Calculations were made with the barrel vibration code for a 1.2 inch constant diameter barrel, which weighs 7.5 pounds. The usual light barrel weighs between 2.7 and 3 pounds. The barrel in the experimental rifle weighs 2.8 pounds. The results indicated that the dispersion error due to vibration would be reduced by roughly a factor of four. Consequently, reducing barrel vibration by increasing barrel weight results in a large weight penalty, although it does work. I once made one of these monsters using a heavy handmade version of the 721 action, and it did shoot well. However, it didn't shoot much better than our experimental rifle modified to eliminate barrel vibration effects. One of the biggest improvements with a heavy barrel is that it is easier to shoot accurately, because it doesn't move around as much as a result of its increased inertia. Someone may wonder if the barrel vibration code could be used to optimize the contour of a light barrel. Well it probably could be used for that purpose, but it would have to be used in a trial and error approach, which is difficult to do. I would prefer to eliminate the vibration in the first place, just as we have done. If the barrel vibration is eliminated, barrel weight is no longer an important consideration in a sporter.

While making barrel vibration calculations I decided to try to determine the most important parameter in a barrel–stiffness or weight? The neat thing about a computer simulation is that you can change both material stiffness and density in a completely arbitrary way and see what happens. In fact in some cases that I ran, the gun would have had to be made of "Unobtainium"! Well, the upshot of all this is that a heavy, flexible barrel is the best. I think this will shock most target shooters, because I'm always reading articles that tell you how to optimize the stiffness in a barrel. People have even milled longitudinal slots in the barrel in an attempt to reduce weight while maintaining stiffness, which is just the opposite to what the computer simulation is telling us. Most of the experts seem to think that fluting doesn't help. Well, if you stop and think about it, if you had a heavy barrel that was hinged at the action so that the torques generated in the forward receiver ring could not be transmitted to the barrel, the barrel wouldn't move much. Unfortunately, this is an impractical solution, so we have to compromise. A possible compromise is to use a light sporter steel barrel, that is as flexible as you can get, and add an overcoat of lead that more than doubles the barrel weight without significantly increasing the stiffness. I tried the lead coating once to see if I could make it work. The muzzle diameter was 0.97 inches at a barrel length of 24 inches, which meets the bench rest rules. It did not work because the lead sleeve came loose. Of course, this is not a useful solution in the case of a sporter, but could be useful in the case of a bench rest rifle if one could get it to work. The sleeve might stay put if the barrel were given a rough finish.

Muzzle Weights

Weights have been attached to the muzzles of rifles in an attempt to improve accuracy, and under the right conditions they probably work. One indication that they may be effective is that according to the literature, the addition of recoil compensators to the muzzle improves accuracy. I have had no first-hand experience with recoil compensators, so I don't know whether or not this is true. The barrel vibration computer code was used to compute the effect of muzzle weight on dispersion, and you can see in Figure 4-37 that the addition of 0.18 pounds to the muzzle reduces dispersion due to vibration by about a factor of 2 on the unmodified, standard rifle. This is a large improvement for such a small weight penalty. One thing that I am sure of with

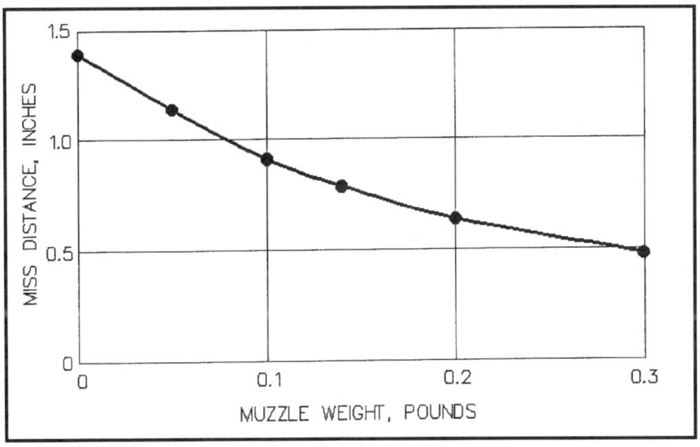

Figure 4-37 - Calculated effect of a muzzle weight on the miss distance due to reduction in barrel vibration. A small muzzle weight theoretically reduces barrel vibration significantly.

regard to muzzle weights is that they must be rigidly attached. Otherwise, they can cause large dispersion. Silver solder seems to be the only reliable method. Unfortunately, I don't know any good way to test a muzzle weight to make sure it is really working. Later, when we have eliminated the other errors that are still present in the rifle there shouldn't be any difference, because most of the barrel vibration should have been removed by the action modifications and the Recoil Isolator. Therefore, we may be able to check again to see how efficient the modifications have been.

Theoretically, a tuned mass damper, which is a spring mass gadget that if tuned properly, could be used to damp muzzle vibration. The problem with using a mass damper is that it will only damp a single frequency and we have several modes present in barrel vibration. So, it gets tricky to apply them. They have been successfully used to damp tall buildings and rotating machinery where the vibration consists primarily of a single frequency. I have tried mass dampers in computer simulations where we have a single third mode and it works, but I don't know how well it would work in practice.

Other Actions

Earlier, it was stated that the Remington 721 was chosen primarily because it has a cylindrical receiver, which makes strain gage instrumentation easier. Many actions, such as the 98 Mauser, Winchester Mod 70, and others, have a flat projection on the bottom of the receiver, which will greatly complicate the strain gage measurement of moment on the forward receiver ring. In fact, I am not sure that such a measurement can be made with any confidence. The other thing about these unsymmetrical receivers is that they would be difficult to modify in order to improve symmetry. The question arises as to whether or not barrel vibration is worse on this type of action as a result of the asymmetry. Since I haven't made measurements on this type of action, I can only guess that barrel vibration would be worse for the same barrel weight and receiver ring thickness. However, most rifles with this style of action seem to have heavier barrels and larger diameter receiver rings. Consequently, it is possible that these differences compensate, at least to some extent for the asymmetry, by adding the extra stiffness and weight. This could be investigated by measuring acceleration on the muzzle. However, I don't intend to pursue this matter, because in view of the results already obtained, the flat bottom receiver appears to be a poor design.

Throat Erosion

Let's take time out to talk about throat erosion, since we have inadvertently stumbled into it. Throat erosion is caused by three known mechanisms.

1) ablation - Shear stresses developed in the moving gas layers next to the bore surface tear away steel particles.

2) chemical erosion - Oxygen molecules and ions chemically combine with iron molecules on the surface of the bore to form iron oxides, which are weak and easily torn away. This process is similar to oxy-acetylene gas cutting and increases with temperature.

3) mechanical erosion - Graphite, primer grit, and unburned carbon particles strike the bore surface mechanically removing steel particles. The bullet jacket scrapes molecules off the surface of the throat.

All of these mechanisms contribute to throat erosion, and a lot of effort has gone into improving powder and barrel steel to reduce the effect. However, it is not clear which one is the most damaging. Forced to make a choice, I would pick chemical erosion as much the worst of the three. It is clear that high pressures and high temperatures increase erosion, and this would be true of all three mechanisms. Also, the larger the case capacity, relative to the bore area, the faster erosion occurs. In fact, cartridges like the 220 Swift and the magnums often will burn out a throat in one to two thousand rounds. However, the 270 Winchester is a standard cartridge and should last at least 3,000 rounds under normal conditions. I think the early and drastic throat erosion in this situation resulted from firing under unusually high temperature conditions. I often fired 40 rounds fairly rapidly without cooling when taking data for this chapter. When all the electronics were working properly, I worked as fast as I could to get the data. Since the ambient temperature was often in the 90's, the barrel became very hot.

The only solution to the problem is to keep the barrel reasonably cool. I normally pour water down the bore through a 1.5 foot long 3/8 inch diameter copper tube which has a plastic funnel attached to the tube with rubber goop, after each two or three five shot groups. This couldn't be done with the instrumented rifle mounted on the machine rest. I don't know whether or not the use of IMR4831 powder contributed to the rapid throat erosion, but it does seem to leave more powder residue in the bore than other powders.

Anyhow, I was surprised to see so much erosion occur in two or three thousand rounds, and while there is no way to prove it, high temperatures were probably the cause.

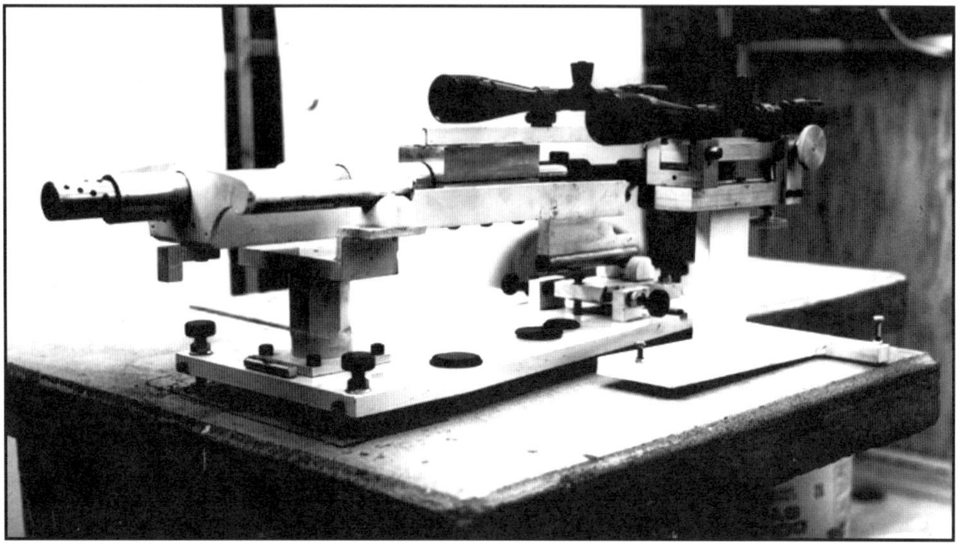

Figure 4-38 - Photograph of rail gun used in testing. See Appendix F for complete description.

Special Bench Rest Gun Problem

Later on during research on muzzle blast effects, I switched from a sporter to a rail gun (Figure 4-38 and Appendix F) and to 6mm cartridges. The switch to 6mm cartridges was made so that I could use bench rest match bullets which are much better than ordinary commercial bullets. I also built a Tunnel Range (Appendix E) with the help of the Zia Rifle Club to eliminate wind effects. The goal was to eliminate all dispersion errors not associated with ammunition defects and was successful. In the process of modifying the rail gun so that it was as free of barrel vibration as I could make it, I measured the moment on the barrel near the barrel block mounts and found a very low amplitude oscillation at 4-5 kc and 9-10 kc. At the time I didn't think these very low amplitude high frequency oscillations could cause a problem, but it turned out that they did.

The rail gun has a 1.350 inch diameter cylindrical barrel clamped to the carriage with solid aluminum blocks, and the muzzle only extends 18 inches beyond the barrel blocks. The carriage, which weighs 45 pounds slides to the rear during recoil on low friction Teflon bearings. This is a very rigid system and as a consequence the barrel vibration frequencies are much higher than a sporter. When I fired groups with different powder loads I noticed that the point of impact changed more than one would expect from differences in gravity drop. So, I fired 3 or 4 five shot groups at each powder load varying from 26 to 30 grains at half grain intervals and measured the average center of impact with respect to a single reference. The muzzle velocity was also measured. Since this gun averages about 0.180 group sizes the accuracy of the vertical group center measurements is fairly good (±0.020 inch). The data were corrected for the varying gravity drop due to varying velocity and plotted in Figure 4-39 compared to a 9.5 kc sine wave. The sine wave is, in effect a representation of the vertical velocity of the muzzle (divided by 10) when the bullet exits. The peaks in the data indicate the maximum deflection in the vertical direction either upward or downward from the mean. Since the flight time at 100 yards is approximately 0.1 second, one can obtain the vertical velocity by dividing the peak value by 0.1. Of course the result is 1.2 inches per second. Now the importance of knowing this is that one can choose a powder load or muzzle velocity that is optimum for reducing this error. Note that if you operate on the peaks (i.e., 26, 27, 28, 29.2 grains) you can

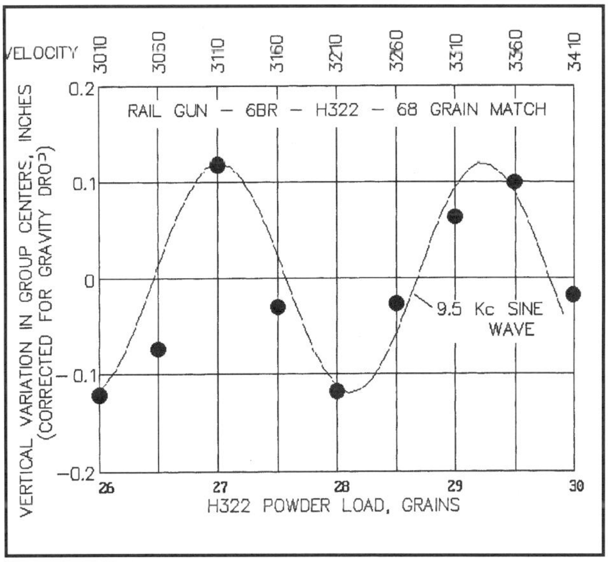

Figure 4-39 - Plot showing variation of vertical position of rail gun center of group impacts at 100 yards at different powder loads and muzzle velocities compared with a 9.5 Kc sine wave. The 9.5 Kc frequency was observed in barrel vibration measurements. The impact points were corrected for differences in gravity drop.

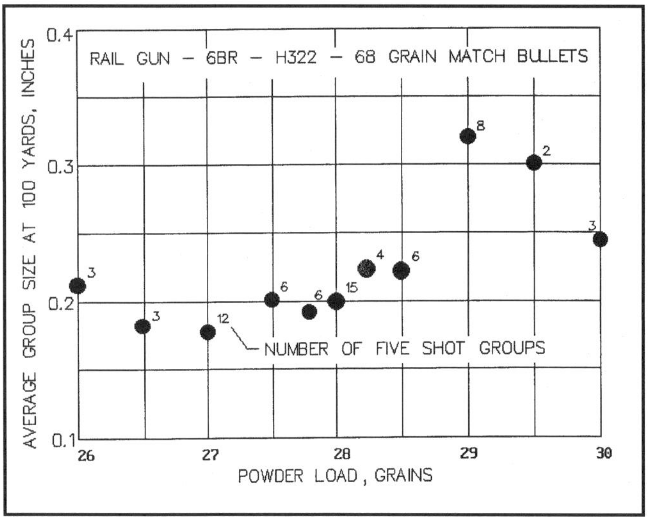

Figure 4-40 - Graph showing 5 shot group size from the rail gun in the Tunnel Range with different powder loads.

have a muzzle velocity extreme spread of 40 fps (equivalent to ± 0.2 grains) without having a large variation in impact due to this effect. On the other hand operating at the crossover points (i.e., 26.5, 27.5, 28.5, 30 grains) where the slope is steep you would expect to see a vertical error contribution of 0.16 inches due to variation in muzzle velocity of ± 20 fps. While the group size data (Figure 4-40) only roughly correlate with Figure 4-39, the best results seem to be at a load of 27 grains of H322 powder. This group averaged in the high one's (i.e., around 0.18 inches). The worst group averages were in the low three's (i.e., around 0.30 inches) at 29 grains. The loads below 27 grains are too light for consistent muzzle velocities.

We found a similar variation in the vertical position of groups with a heavy varmint 6PPC bench rest gun belonging to a friend (Dr. Jackson). Figure 4-41 shows a plot of the vertical variation of point of impact with gravity variation removed for this gun. Notice that the frequency is lower as one might expect—about 6.7 kc instead of 9.5 kc. But it appears to be the same phenomenon. The vertical stringing of the group size (not shown) is at a minimum below 3100 fps and above 3300 fps. This correlates with the negative slope of the sine wave, and I believe this is to be expected. If the muzzle velocity is higher than the mean for the group it would have a little less gravity drop and normally would impact a little higher. But, if the gun is operating on the negative slope the higher shot will be corrected downward to the group center by this high frequency vibration phenomenon. The best place to shoot is just past a positive peak on the curve. We have found that

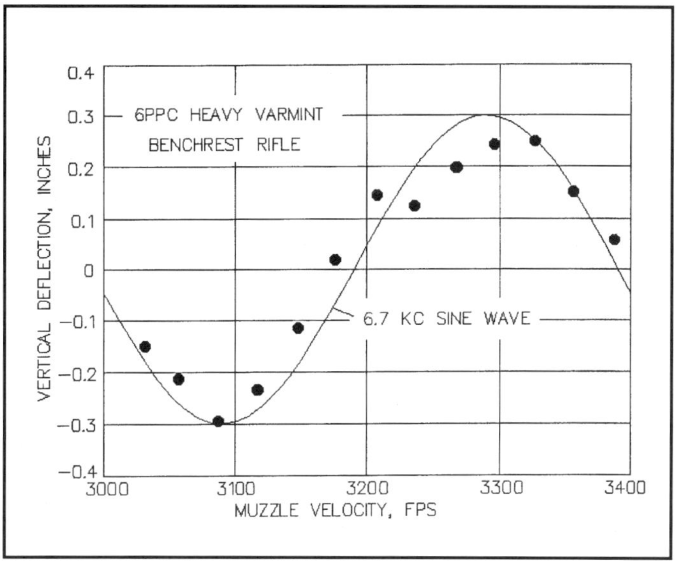

Figure 4-41 - Plot of the vertical positions of group centers for various muzzle velocities on a 6PPC benchrest rifle. The effect of varying gravity drop due to differences in velocity has been removed.

the amplitude of the sine curve is less (about half) with the free recoil method of shooting than it is with the firm hold method. This is not surprising, since some of the shooters body weight is effectively transmitted to the stock in the firm hold approach. This increases barrel vibration amplitude. In the free hold only the trigger is touched by the shooter. Of course, you can't use the free recoil method with light sporters or heavy recoiling guns. Consequently, this only applies to heavy bench rest guns.

I decided to test my bench gun which is the completely modified action including the Recoil Isolator with a heavy varmint Shilen barrel chambered in 6mm BR. The data are shown in Figure 4-42. You can see that when the data are corrected for the variation in gravity drop it becomes essentially a horizontal straight line. This shows that there is very little barrel vibration involved in this gun, otherwise we would see a sine wave variation in group center of impacts in the vertical plane just like we saw in the rail gun and the custom heavy varmint gun.

While this is not a problem with sporters, it could be a problem with bench rest guns where shooters try to shoot groups that average less than 0.2 inches at 100 yards. Vertical stringing of groups is common in bench rest guns and the typical approach is to keep increasing the load until it stops. Unfortunately, this won't always work because you run into the maximum pressure restriction or the vertical dispersion may be caused by another problem. Vertical stringing of groups can be caused by bolt lugs not seating evenly on

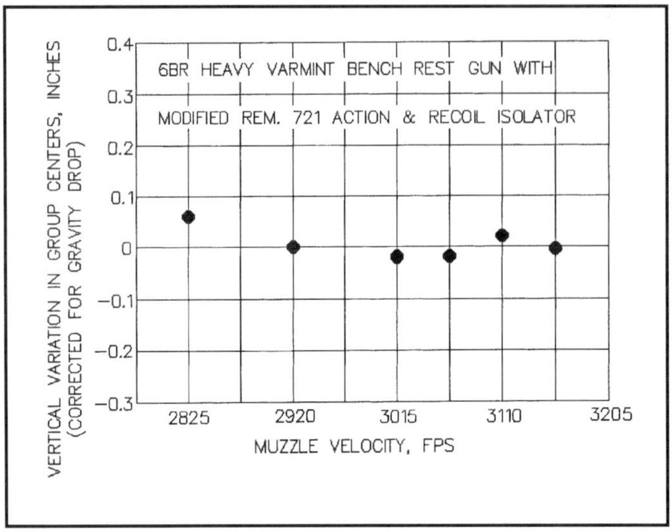

Figure 4-42 - Plot of the vertical positions of 5 shot group centers for various loads and muzzle velocities in the modified 721 action with recoil isolator and heavy varmint 6mm BR barrel. Gravity drop variation is removed.

the receiver lugs. In a rail gun this bolt lug problem is eliminated. With a rail gun you can change the vibration frequency by moving the barrel either forward or backward in the barrel blocks, which changes the phase of the oscillation. This may allow one to find a reasonable load that minimizes vertical dispersion. With a bag gun (i.e., bench rest gun fired from sand bag rests) one might reduce the length of the barrel to obtain an optimum result, although I would check the top and bottom bolt lugs for equal bearing first. I would also be suspicious of the threaded barrel joint in a bench rest gun (See Chapter 6).

Once you have established the best velocity for accuracy you should test every new bottle of powder to make sure that you are getting the same velocity for a given load. We have found that different bottles of powder with the same lot number sometimes have different characteristics.

Summary

We have at this point measured the moments on the forward receiver ring and evaluated their effect on accuracy. We have also eliminated the recoil moment through the use of a special bedding device called a Recoil Isolator, which does not transmit the recoil force from the stock to the receiver until the bullet has left the barrel. This eliminates the recoil moment effect on the receiver that contributes to barrel vibration and inaccuracy. We also found that the receiver structural asymmetries were another source of moment and made modifications to the receiver to eliminate these sources of vibration. The motion of the muzzle of the barrel was measured with an accelerometer, and these data proved that we had greatly reduced barrel vibration.

A barrel vibration computer simulation code was used to estimate the contribution of barrel vibration to the dispersion of the rifle, and found it to be about 0.84 inches on the standard unmodified rifle. Accuracy tests were run with five shot groups at 100 yards, which show that the normal average group size of 1.5 inches on a standard rifle was reduced to 0.884 by the modifications that were made.

We also demonstrated that barrel vibration causes a vertical shifting of the center of impact of groups with changing muzzle velocity.

Now that we have effectively eliminated barrel vibration, there are at least six more significant errors remaining in the rifle that have to be corrected. They are scope sight motion, barrel joint motion, muzzle blast effects, bullet core problems, bullet imbalance, and external ballistics problems. We work on scope sight problems in the next chapter.

Congratulations! You have just made it through the most difficult part of the book. I promise that there will be no more electronics and other stuff that makes for difficult reading.

CHAPTER 5
SCOPE SIGHT PROBLEMS

Scope sights and their mounts have mechanical problems that can cause dispersion, so we will take care of this before we get into some of the more complicated and less obvious problems. These mechanical problems generally fall into two categories, motion of the optics and motion of the mounts.

Optical Parts Motion

A number of years ago I bought two expensive high-power variable scopes of a well known brand that were identical. I noticed that my shooting accuracy suddenly deteriorated, and decided something had to be wrong with the scopes. The only thing to do was to mount the receiver in a rigid vise then jar the mounted scope and see if the reticule returned to the same aiming spot. Well you guessed it. Every time I gently tapped either one of these scopes, the reticule returned to a different spot! I repeated the experiment with another scope of different manufacture and the reticule always returned to the same spot. There was just no doubt about it, some of the optics inside the scope were not rigidly mounted. The problem turned out to be in the way the objective lens (i.e. front lens) cell was mounted in the parallax adjustment mechanism. Well, I fixed one of these scopes by modifying the objective lens cell mount and traded the other one off for another scope. I won't mention the brand that was unsatisfactory, because it happened too many years ago and they may be perfectly satisfactory by now. This experience taught me that scopes can be faulty, and the only way to tell is to bench test them. When you run a test like this you should use a heavy, rigidly mounted vise with lead lined jaws. I use a small, light piece of softwood to tap the scope, and you need to tap the scope in several places and at different angles. It doesn't take much of a blow to make the reticule jump, and you don't want to hit the scope tube too hard or you might dent it. So far, I have not had trouble with Weaver, the old Leupold 20X, or Bausch and Lomb. Recently I have had trouble with a new 24X target scope, and after making several telephone calls I found that everybody in the bench rest fraternity was having the same problem. A few people have started small businesses rebuilding some target scopes. Now to put this in the proper perspective I have to tell you that the recoil acceleration is much higher than normal on our experimental rifle, as the following table shows.

TABLE 5

Table 5 - Peak Recoil Acceleration

Rifle	Recoil Weight	Acceleration (g's)
Experimental 270	6.25	480
Standard 270 Sporter	8.25	363
Light Varmint, 6PPC, 6BR	10.5	216
Heavy Varmint, 6PPC, 6BR	13.5	162

Well, you can see that a scope gets a very severe ride on the experimental rifle because the recoil isolator causes the recoil weight to be lower than other rifles with the same total weight.

Also, note that the acceleration is much lower on typical bench rest rifles. This means that a particular target scope may be all right on a heavier rifle, without the recoil isolator.

There is one way around this problem that may eventually work, and that is to use a spring loaded sliding scope mount, similar to the old Unertl mount. I tried an old Unertl scope and mount, but it was never designed to take these heavy loads and did not work. I then tried designing a mount using a similar approach, but it also failed. In retrospect, this mount had several design flaws and could not have performed properly. However, I believe that the sliding mount approach could be used to successfully reduce the recoil acceleration, but it will take a lot of work.

This scope optics movement problem is very insidious, because it is difficult to detect. If I hadn't been shooting a very accurate rifle, I might never have noticed that the scopes were defective. The only way to find out is to test by both bench testing and firing. In spite of everything you do I don't know of any way of being absolutely certain that the scope optics are not moving.

Scope Mount Motion

I have always been suspicious of scope mounts. These things take a heck of a beating on a high powered rifle, and I have never been certain that they stayed put. The mounts on this particular rifle are Weaver Top Mounts with aluminum bases attached by two 6-48 screws. Now, there is just no way that two small screws can keep these bases rigidly fixed to the receiver under the loading conditions present on a rifle, no matter how tight you get them. This becomes very evident when one realizes that the axial load on the scope is roughly 500 to 700 pounds on a 270 sporter during firing. Well, in fact the bases don't stay put, and you can prove this with a very simple test. All you have to do is tap on the front base with a small hammer applied to a wood dowel so as to move it to the right, and repeat the operation on the rear base in a direction to move it to the left. Then you fire a shot, and repeat the operation three times for a three shot group. The whole operation is repeated again, only this time the bases are tapped in the opposite direction. If there is no effect, all six shots should be grouped together. When I ran this test, two distinct three shot groups resulted separated by 0.503 inches in the horizontal

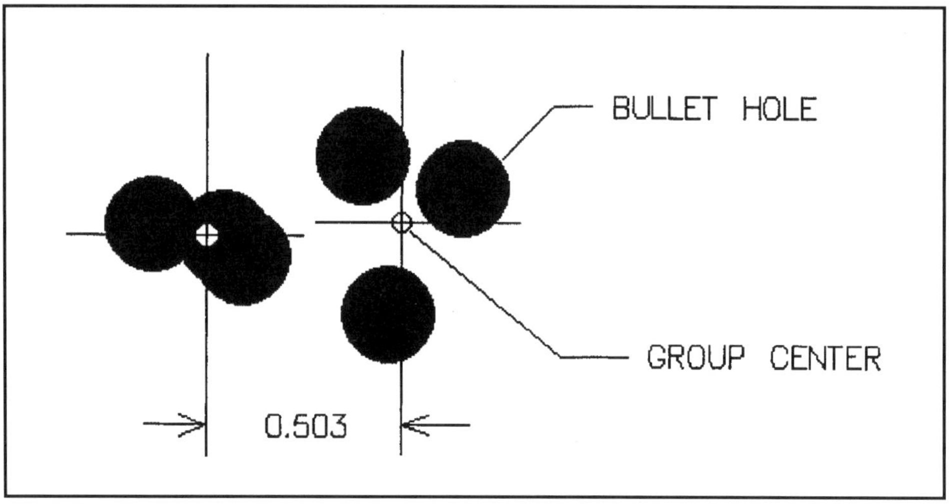

Figure 5-1 - Computer representation of an actual target showing two distinct groups caused by tapping the sight bases to skew the scope sight mount bases first to the left and then to the right.

direction. A computer representation of this target is shown in Figure 5-1. In order to get this large an error, the two bases need only move by ±0.18 mils. Not very much motion, and about the amount of motion that might be expected in a small screw. By the way, the screws were tightened by impact driving, which is the only way to get them really tight. I should also mention that I have tried several types of chemical bonding between the bases and the receiver, but they always shot loose. Of course, this problem is not the fault of the mount, but is the result of the way the receiver is designed to accept scope sight mounts. Some of the newer bolt action rifles have taken this into consideration, and have grooves milled into the receiver to accept a clamp-on mount. Hopefully, this design change should solve the problem. However, we are stuck with this problem on this particular receiver, and we want to eliminate it completely so that we can get on with our investigation of other problems. Consequently, I made copies of the aluminum bases using steel, and I used a low temperature (i.e. 430°F) silver solder to attach them to the receiver. Don't confuse this stuff with ordinary solder, because it is much stronger, having a shear strength that can approach that of mild steel (i.e. 15,000 psi). Just to be sure of the strength of the silver solder, I decided to test two different types of joints in a calibrated hydraulic press. The first was a lap joint, which had a shear strength of 4500 psi, and the second was a cylinder in a matching hole, which had a shear strength of about 14,000 psi.

The scope sight base joint should have a strength that lies somewhere in between these values, because the two test cases represent the extremes in joint strength. Based on the hydraulic press test results, the minimum strength of the joint between the receiver and the bases should be about 8,000 pounds, which means that the silver soldered joint should withstand more than ten times the actual load. In fact, I test them by trying to knock them off with a hammer and a brass punch to make sure of the bond. This modification should only be made by a skilled gunsmith, because of the obvious possibility of altering the heat treatment and the strength characteristics of the receiver. After firing the rifle a number of times with silver soldered bases, the test depicted in Figure 5-1 was repeated, and there was no evidence of motion of the bases. Now before someone notices the small size of the two three shot groups and decides that there is no point in going any further, the reader is advised that this test was run later after several other corrections were made to the experimental rifle.

At this point in the experimental investigation I decided to build another rifle exactly the same as the 270 with silver soldered steel bases but chambered for 6mm Remington. The reason for doing this was that high quality match bullets were not available for the 270 but were available in 6mm. I knew that the ordinary 270 bullets that I was using were poorly balanced and contributed a large error. It seemed like a good idea to try to minimize the bullet problem, at least for the time being, to help isolate some of the other problems. This turned out to be a good move because the first set of groups revealed another scope mount problem. Figure 5-2 shows two groups fired

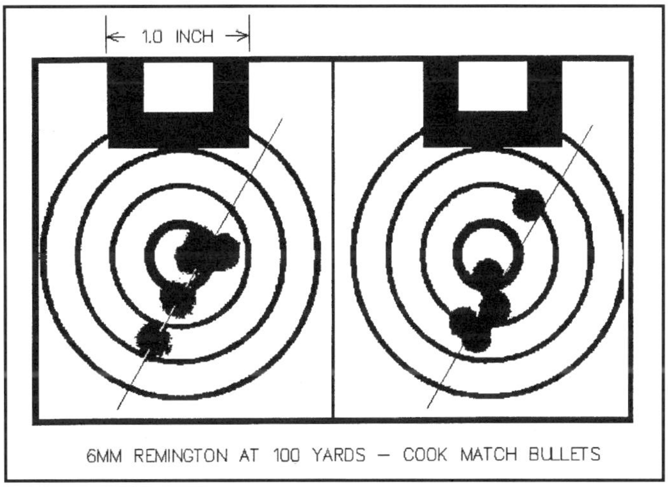

Figure 5-2 - Typical 6mm Remington targets fired with Cook match bullets showing vertical dispersion due to the axial load developed by the scope between the front and rear scope mounts.

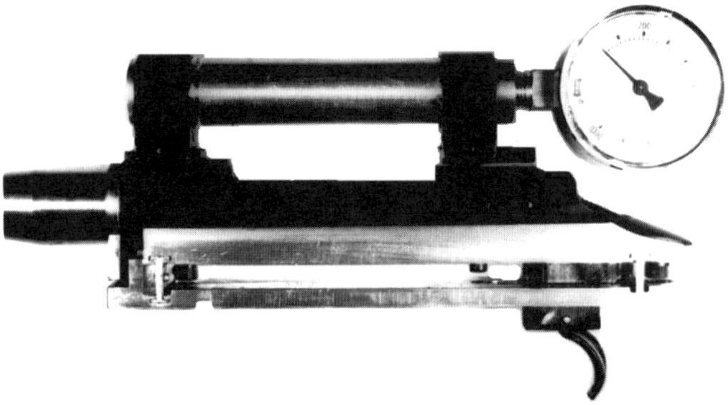

Figure 5-3 - Photograph of the hydraulic cyclinder used to impose a compression force of up to 200 pounds between the front and rear mounts. Muzzle deflection was measured by dial gages.

with the 6mm using Cook 65 grain match bullets. You can see that the bullet holes are scattered in a string vertically and to the right. While a lot of things could cause this type of dispersion, it turned out to be differential axial motion of the scope between the front and rear rings. This causes a differential compression or tension axial force to develop between the two rings. According to a theoretical calculation the force could approach 200 pounds. This force warps the receiver causing the barrel to point in a different direction. Just to prove this theory I made a closed hydraulic cylinder with a pressure gage attached that replaced the scope (Figure 5-3). The action was held in a vise and a dial gage measured the deflection of the muzzle as the hydraulic fluid was heated causing a axial force between the front and rear scope mounts. The results are tabulated in the following table.

TABLE 6

Effect of Differential Force Between Scope Mounts

Force	Muzzle Deflection (mils)		Miss Distance @ 100 Yards(in)	
Pounds	Vertical	Horizontal	Vertical	Horizontal
176	12.0	4.5	1.66	0.62
88	6.0	2.5	0.83	0.35
44	3.0	1.0	0.42	0.14

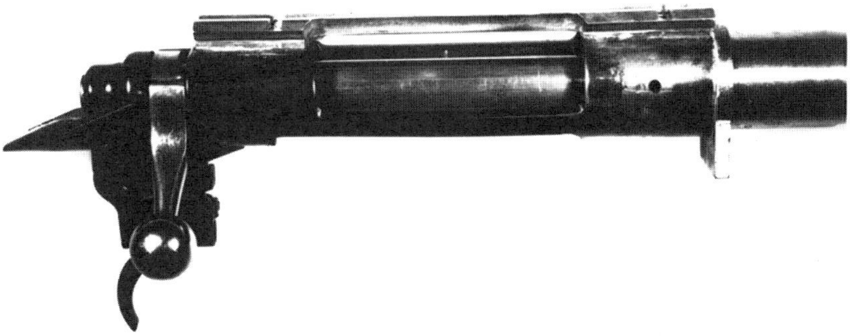

Figure 5-4 - Photograph of the steel bridge mount on the rifle action, which was used to solve the scope mount axial differential loading problem.

From the data you can calculate the angle with respect to the vertical to be about 23°, which is a little less than the angle seen on the targets. The reason for the groups being canted at an angle is that the receiver is weaker on the right side than the left as a result of the loading port. Consequently, the receiver bends in a plane canted to the right with respect to vertical. One can also see from Table 6 that a differential force of roughly 75 to 100 pounds is all that is required to cause the amount of linear dispersion seen in Figure 5-2. I decided that the easiest way to solve this problem was to make a steel bridge mount and silver solder it at both ends (Figure 5- 4). The bridge mount worked and eliminated the problem. I repeated the test in Table 6 with the bridge mount and the muzzle deflections were reduced by roughly a factor of ten, which means that the vertical dispersion caused by the scope mount differential axial load should be less than 0.1 inch at 100 yards. The bad part about this solution is that it interferes with the loading port and is a bit of a nuisance. It also adds about 1.5 ounces of weight, but I don't know of any other solution. However, the bridge mount does add considerable stiffness to the fairly flexible receiver, and should improve accuracy in other ways besides the scope problem. To further reduce the scope differential loading problem, I lubricated the clamp and saddle of the rear mount with a teflon lubricant (i.e. Friction Block) and didn't tighten the screws quite as tight as I normally would. Whether or not this helped is not known, because the effect on group size was too small to be detected.

There is one other source of scope sight motion that I discovered later in this work, and that is motion of the scope tube in the mounting rings. The bottom part of the scope mount that clamps onto the bases has a circular cradle designed to fit the scope tube. Unfortunately, the diameter of the circular cradle is about 5 mils too large, and the clamping band is too wide, so the scope tube is really only restrained in the horizontal direction by friction (see Figure 5-5). You might think that the clamping bands would distort the scope tube enough so that the tube would conform to the cradle, but this is not the case. The scope tube only deforms about one mil on the diameter when the bands are very tight. If the scope moves in one of the mounts just 0.15 mils in a horizontal direction, the shot will be displaced by 0.1 inch at 100 yards. While I was never able to prove conclusively that the scope tube was moving in the mounts, there were strong indications that it was moving and contributed as much as 0.3 inches of horizontal dispersion at 100 yards. The reason that this effect is difficult to prove is that wind effects get into the act, making it difficult to be absolutely sure. You see, I didn't have the tunnel range at this time. Fortunately, it is an easy problem to fix, as you can see in the drawing shown in Figure 5-5. All we have to do is scoop out the bottom of the circular cradle with a 3/4 inch ball end mill and the scope tube is then forced to contact the inside of the mounting ring at three points that are equally spaced. The amount of material removed by the ball end mill (30 mils) is exaggerated in the Figure 5-5 for the sake of clarity. Another method is to bed the scope tube in the bottom part of the mount with Devcon F aluminum filled epoxy. I use black shoe polish instead of using other release agents, and hold the scope

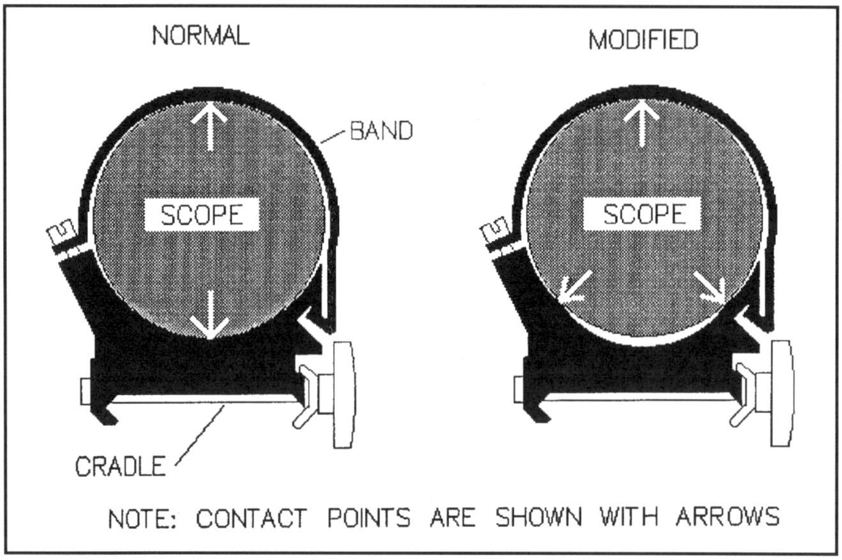

Figure 5-5 - Sketch of a Weaver top mount demonstrating a modification to the cradle to eliminate the two-point contact with the scope tube.

onto the mounts with strong rubber bands until the epoxy hardens. Either approach, or both used together, will prevent the scope from moving in the mounts, and we can quit worrying about it. Other types of scope mounts may have this same problem. However, I doubt that this small effect can be observed in an ordinary sporter.

Parallax

Parallax is easily detected by moving the eye in a lateral direction behind a fixed scope, and observing motion of the reticule with respect to the target. It is a fact of optical principle that parallax can only be perfectly eliminated at one range for a given adjustment of a telescopic sight, and the higher the power the more critical the adjustment. Low powered scopes usually can't be adjusted for parallax and are designed to be free of parallax at some average range. High power (i.e. more than 9X) scopes have a special adjustment ring that is usually calibrated for range. Unfortunately, I can never seem to completely eliminate parallax in most high powered scopes (i.e. 20x to 36x) with the adjustment provided. Maybe you don't have this problem, but if you do, you can eliminate it by moving the eye back beyond the optimum eye relief and centering the circle that appears in the scope field of view. This approach has the effect of keeping your eye centered on the optical axis so that parallax doesn't matter. This has an additional advantage of reducing the chance of eye contact with the scope during recoil.

Optical Refraction

Optical refraction, which is more often referred to as mirage, occurs when there are small but significant density changes in the air space between the shooter and the target. These density gradients cause the light rays to be bent and distorted. As a result, the target image appears to move and may be distorted or blurred. Wind and high ambient temperatures usually make mirage worse, although, mirage can be severe under cold conditions. Mirage is usually worse close to the ground, and it is particularly bad over bare ground. The only really satisfactory way to deal with this problem is to pack up and go home and come back another day. However, we don't always have that option, and with experience you can minimize the effect.

There are really two types of mirage. One type is what I call shaky mirage which is rapidly changing and the other is a very slowly varying type of mirage that most shooters are unaware of. The shaky stuff is often caused by the hot air rising off a warm barrel right in front of the scope sight, although it can occur on a hot windy day by hot air boiling off the ground. If the shaky stuff is being generated by the barrel you can attach a piece of thin plastic about 3 or 4 inches wide over the barrel in front of the scope which deflects the warm air to the sides. Velcro is usually used for attaching the plastic sheet to the barrel and works well. The long tubes that screw into the objective lens cell of target scopes may help but can cause movement of the lens cell. For this reason most target shooters no longer use them. If the shaky mirage is caused by hot air beyond the muzzle there is another way to compensate. If you watch closely, you will be able to notice the occasional, momentary appearance of what appears to be a clear, well focused image. Well the trick is to get lined up on this real image and squeeze the trigger when the sight is lined up properly during one of these opportunities.

However, other shooters use a different approach, and use the mirage as a wind indicator. The idea is to line up the cross-hairs on the clear image when the wind clears away the mirage and then fire when the mirage boils up and obscures the target. The theory behind this is that the cross wind is at a minimum when the mirage is at a maximum. This assumes that the gun doesn't move while you are waiting for the mirage to boil, so it can't work for anything but bench rest shooting. The only other solution that I know of, and it isn't very practical, is to go shooting on the moon where there isn't any atmosphere!

The slow type of mirage is difficult to detect but it is present on open ranges. I built a mirage reference scope adjustable mount that holds a 36X target scope on the bench (Figure F-1, Appendix F). By watching the target through this scope you can tell if slow mirage is present and correct for it. In the Tunnel Range slow mirage can cause a drift in the vertical direction of 0.6 inches without the exhaust fan. The exhaust fan essentially eliminates this problem if you match the outside air temperature with the tunnel wall temperature.

Optical Resolution

Optical resolution is one of the limitations on our ability to aim a rifle at a target. It depends on magnification, scope optics, atmospheric conditions and the diameter of the objective lens. Of course, it also depends on an individual's visual acuity. Unfortunately, we don't have a lot of control over most of these factors. While it appears to me that optical quality depends to some extent on price, most scopes these days have excellent optics. The diameter of the objective lens is fixed by the largest diameter objective tube that you can hang on a rifle without it becoming awkward and ungainly. Most spotting scopes have a larger objective than a rifle scope has, and you can tell the difference in resolution, at the same magnification, by simply looking at the same scene and comparing them directly. The diameter of the exit pupil, which is the diameter of the column of light that comes out of the eyepiece and enters the eye, is equal to the diameter of the objective lens divided by the magnification (i.e. power). This means that the higher the magnification the smaller the exit pupil becomes. Ideally, the scope exit pupil diameter should match the diameter of the pupil of the eye for maximum illumination. In dim light, the pupil of the eye may have a diameter of 5 mm or larger, while in bright light it may shrink down to 1 or 2 mm. Consequently, hunting scopes are generally designed to have an exit pupil of around 5mm, while target scopes usually have an exit pupil of around 2 mm. An example is the Leupold 20X shown on the rifle in Figure 2-1. This scope has an objective lens diameter of 40 mm, and consequently, at a power of 20 has an exit pupil of 2 mm. What all this amounts to is that you can improve resolution by increasing magnification up to the point where the exit pupil becomes too small for the lighting conditions.

Determining the optical resolution of a scope is an "iffy" proposition, and depends a lot on viewing conditions, but my best guess is that it is around 10 mils at 100 yards with a 36 power scope under good conditions. It also seems to me that resolution is roughly proportional to magnification, as long as the exit pupil remains about the same. This means that our aiming accuracy can be no better than maybe 10 mils, and is enlarged by the things that we have previously discussed. Of course, another factor is the visual acuity of the individuals eyeball. Shooting accuracy, which is the accuracy with which one can aim the rifle and release the trigger without disturbing the alignment of the rifle, is discussed later in the book (Chapter 11).

CHAPTER 6 BARREL-RECEIVER THREADED JOINT MOTION

It would hardly seem possible that the threaded barrel joint could move, causing the barrel to point in a slightly different direction after a shot is fired. But, that is exactly what happens, and barrel joint motion can cause large flyers (i.e., one inch or more) in a group. The only things that prevent motion of the barrel joint is the lateral friction forces caused by the axial preload that results from the applied torque on the barrel when it is installed and the stabilizing forces acting on the angular surfaces of the threads. Unfortunately, applied loads (e.g., bolt thrust during firing), and differential thermal expansion can either reduce the axial preload or completely overwhelm the stabilizing effect of the axial preload. Under some temperature conditions, such as rapid fire, it is possible for the joint to be completely unloaded or loose when the gun is fired. So how do we know that the joint is moving? We make some measurements.

Barrel Joint Motion Measurement

All of the experimental measurements will be made on a Remington 721 action with a barrel chambered for the 270 Winchester cartridge. This action has a 1.0625X16 thread. What we need is some simple way to determine which way the barrel is pointing with respect to the receiver after each shot.

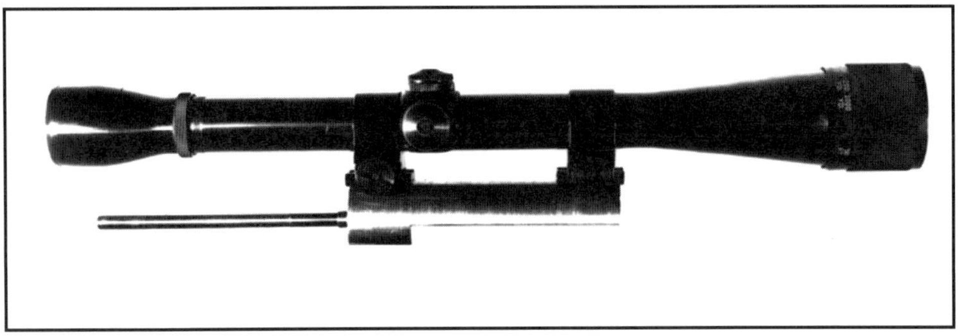

Figure 6-1 - Photograph showing Weaver K12 scope mounted on a mandril that slips into muzzle. By comparing aiming point on the target of the muzzle scope with the regular scope mounted on the receiver, motion of the barrel joint between shots can be detected.

Well, the simplest way that I can think of is to attach a second scope sight to a mandril that slips into the bore, which tells us where the barrel is pointing. The rifle has to be mounted on a machine rest for this experiment. The test routine is to adjust the machine rest so that the receiver scope is pointing at the aiming point, and then insert the muzzle scope in the muzzle and adjust it so that the cross hairs are also on the aiming point. After the shot is fired the machine rest is checked to make sure that the receiver scope is still pointed at the aiming point, and then the muzzle scope is inserted and the position of the cross hairs relative to the aiming point recorded. In this way you can tell where the barrel is pointing with respect to the receiver. The muzzle scope fixture is shown in Figure 6-1 with the 12 power scope attached. The radial clearance between the mandril and the bore is 0.1 mils, which is a fairly tight fit. With this clearance the pointing error of the scope is $\pm$0.3 inches at 100 yards. So, while this is a simple approach that works, it is also a little crude. A series of four 5 shot groups were fired. The data from the muzzle scope confirmed that when a significant flyer appeared, the barrel was pointing in a new direction relative to the receiver after the previous shot. The next shot would then be a flyer. The barrel would occasionally stay put after firing a shot, but it would move fairly often and return to the original position after firing the next shot which would be a flyer. As you would expect, significant barrel motion was not always observed from shot to shot because the motion would be too small to detect due to the imprecision of this measurement method ($\pm$ 0.3 inches). The magnitude and direction of the shift in barrel position was in general agreement with the position of the bullet impacts. I can't think of any sensible way to present the data in graphical form, so you will just have to take my word for it that the data proved that the barrel joint was moving. A more precise way to perform this experiment might be to permanently attach a laser to the bottom of the barrel near the muzzle. Then one could observe the position of the laser spot on the target compared to the bullet impact as the rifle was fired.

I ran another simple experiment to prove that there is relative motion in the threaded joint between the receiver and the barrel. I had previously noted that the first few shots from a newly installed barrel were always wild. These first shots were much too far out of the normal group to be caused by a clean barrel. It would seem that this would have to be due to the barrel shifting position in the joint at the first shot. I removed the barrel and applied Permanent Loctite, which is a thread locking material, to the threads and replaced the barrel. When I test fired the rifle the first rounds were not wild, and the 5 shot group measured 0.626 inches at 100 yards. After a few more rounds were fired, fliers started appearing, indicating that the Loctite was no longer able to constrain the barrel joint. Epoxy is not strong enough to take the repeated stress of firing, particularly at elevated temperatures. This experiment also proved to me that the barrel joint can move.

The upshot of all this is that I think we can safely assume that there is relative angular motion between the barrel and receiver. We are going to measure the axial preload that is actually applied to the joint when it is tightened, so that we can see if the joint is tight enough to withstand the loads caused by firing.

Barrel Joint Axial Preload Measurements

Fortunately, it is fairly simple to measure the joint axial preload with a strain gage. We do it just like we did when we measured the bolt thrust, except the gage is placed over the threaded portion of the forward receiver ring. With ordinary 10-30 motor oil as a lubricant, the axial preload measured 10,600 pounds when the barrel was tightened with an applied torque of about 250 ft-lbs. This value is consistent with the calculated value of 10,506 pounds for a friction coefficient of 0.24, and an applied torque of 250 foot-pounds. The friction coefficient is a number that when multiplied by the axial preload yields the lateral friction force. Once the lateral friction force is known you can determine the torque required for a given axial preload. The calculated value for the axial preload can be obtained from a textbook equation. The equation for calculating axial preload is

$$F = (12*T)/(P/(2*\pi) + f*Rt/\cos\beta + f*Rb)$$

where

F = axial preload, pounds
P = thread pitch, inches, (1/16)
π = 3.14159
f = friction coefficient, pounds/pound, (0.24)
Rt = average radius of the barrel threads, inches, (0.505)
Rb = average radius of barrel shoulder, inches, (0.565)
β = thread angle, 30 degrees on standard threads
T = applied torque, foot-pounds

The friction coefficient of 0.24 is consistent with handbook (Handbook of Physics and Chemistry) values for petroleum lubricants, although it is only approximate. The friction coefficient also depends on the surface finish which is difficult to evaluate. In this case the lateral friction force acting on the

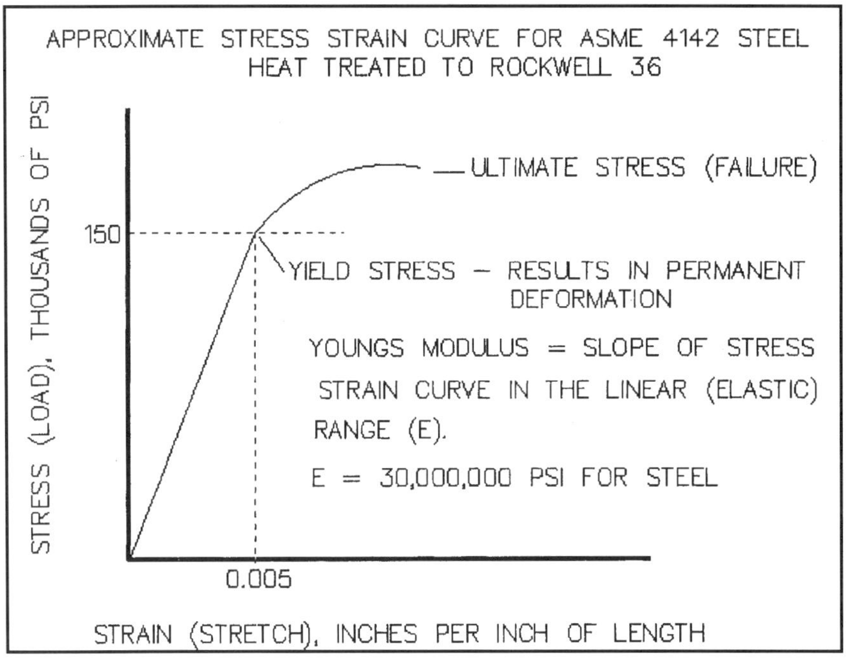

Figure 6-2 - Graph showing how stress (load) is related to strain (stretch) for steel.

TABLE 7

Barrel-Action Joint Tests

Test	Lubrication		Axial	Preload	Results
	Threads	Shoulder	Initial	Final	
1	oil	oil	10,600	10,600	Large flyers under all cartridge case and temperature conditions.
2	Teflon tape	Teflon tape	27,600	19,100	First two groups were excellent, but recoil lug failed in compression and remaining groups were bad.
3	oil	oil	20,300	20,900	Accuracy excellent. Axial preload stayed the same within the accuracy of measurement.
4	oil	oil	24,700	21,900	Good accuracy. Axial preload reduced during firing.
5	rosin	lanolin	17,800	17,800	Very poor accuracy.

threads would be 10,600 times 0.24, or 2,544 pounds. So, we have a measured value for the threaded joint axial preload that is consistent with theoretical calculations. This value of 10,600 pounds is probably representative of the axial preload found on most sporters. However, I did measure the axial preload on one standard Remington 721 with strain gages when the barrel was removed and measured a value of 8054 pounds. Well 10,600 pounds seems like a lot, and you might think that a load that large should freeze the barrel joint. It probably would if it weren't for the other effects during firing that reduce the axial preload and friction forces acting on the joint.

Barrel Joint Tests

In order to determine the effect of different joint conditions, a number of experimental tests were run to determine the effect of the friction coefficient and axial preload. These tests on the 270 sporter are summarized in Table 7.

Some yielding of the first two barrel threads was observed on tests 2 and 4. In order to explain the terms yield and failure a graph is shown in Figure 6-2. Visualize a steel bar that is being stretched by equal tensile forces on each end which produces a stress in pounds/square inch. Stress is shown on the vertical axis. The amount of stretch of the bar or strain in inches per inch of bar length is shown on the horizontal axis. Steel is an elastic material and the stress is proportional to the strain until the steel reaches the elastic limit and starts to yield. The bar will continue to support a higher stress for a while until it reaches the ultimate stress and fails (breaks). One should be careful about overloading the threads and barrel shoulder when using Teflon tape or lanolin with the standard V type threads. Standard V type threads won't support an axial preload in excess of 20,000 pounds before they start yielding.

The approximate friction coefficients for the lubricating materials for steel on steel are

Teflon ———	0.09
lanolin ———	0.10
10-30 oil ——	0.20-0.24
none ————	0.58
rosin ————	>1

Several facts can be determined from the test results. It can be seen from tests 2, 3, and 4 that the joint can't sustain an axial preload above about 20,000 pounds without failing. Another conclusion can be reached, and that is, an axial preload of at least 25,000 pounds on the joint is required to stabilize the joint under all conditions when the joint is lubricated with a low friction coefficient lubricant. However, tests 2 and 4 demonstrate that an axial preload of 25,000 pounds cannot be sustained by the joint as presently designed.

Now, there are two things that stabilize a given threaded joint, the axial preload and the lateral friction force. Unfortunately, it is difficult to determine the relative importance of these two effects. So, at this point (Test 5), I decided to increase the lateral friction force by using rosin as a thread lubricant, and I installed a recoil lug that was heat treated to 130,000 psi yield strength. The new recoil lug was necessary because the factory recoil lug is not strong enough to withstand these large preloads. By the way, the action is factory heat treated to a yield strength of about 190,000 psi, and the barrel is factory heat treated to about 130,000 psi. I intended to apply an axial preload of about 20,000 pounds, but could only reach 17,800 before breaking the action wrench, which was made of mild steel. Just how large a friction coefficient rosin has I don't know, but it must be a lot greater than the friction coefficient of dry steel on steel (0.58), because it is used on the jaws of the barrel vise to keep the barrel from rotating during barrel installation. Lanolin was used on the barrel shoulder to reduce the torsional load imposed during tightening the barrel. The result of Test 5 was that the accuracy was very poor, which indicates to me that the main force that stabilizes the joint is the axial preload, and that the lateral friction force plays a secondary role in keeping the joint rigid. However, the axial preload of 25,000 pounds required to stabilize the joint is more than the standard V threads will take without yielding. At this point I decided to stop and analyze the loads acting on the joint, because this is a real design dilemma.

Barrel Joint Loading

There are several loads acting on the barrel joint that can reduce the axial preload. One of these effects is the differential heating between the barrel and the receiver ring, which can be substantial. By using a thermistor, I measured a temperature difference of 56°F between the inside of the chamber (133°F) and the outside of the receiver (77°F) after firing 15 rounds in rapid succession. One can calculate that this temperature difference between the barrel and receiver will cause enough differential expansion in the axial direction to reduce the axial preload on the joint by roughly 7,000 pounds. Radial expansion of the barrel reduces this effect by some small amount that I don't know how to estimate, so we will stick to the 7,000 pounds. There is no doubt in my mind that an average temperature difference of at least 56°F

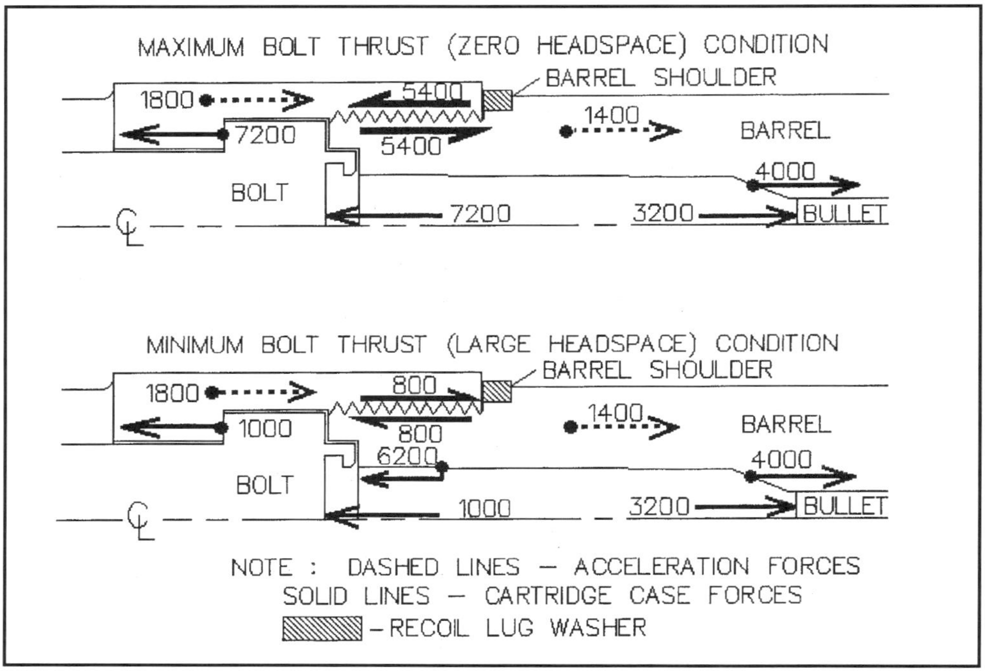

Figure 6-3 - Drawing showing inertial forces due to recoil motion of the barreled action and force due to bolt thrust acting on the barrel joint.

can and likely does occur during rapid firing. There is so much time lag involved in making the temperature measurements that the temperature differential is probably greater. While you are waiting for the thermocouple used to measure the local temperature to reach equilibrium the barrel and action temperatures are gradually equilibrating. Consequently, the measured differential temperature between barrel and receiver will be less than the actual difference in temperature.

Another thing that effects the load on the joint is the action of the cartridge case during firing. We know from Chapter 4 that the bolt force can vary from 1,000 to 7,200 pounds. This is due to the variation in the headspace and cartridge case conditions, causing a difference in the load on the joint. The diagrams shown in Figure 6-3 show the component forces acting on the joint for the two extreme cartridge case conditions of maximum and minimum bolt force. The dashed lines show the magnitude of the acceleration forces and the solid lines show the component forces resulting from the cartridge case. The heavy solid vectors show the direction and magnitude of the resultant forces acting on the threads. The reader should note that all applied forces are opposed by a reaction force caused by the recoil acceleration. I have called these reaction forces acceleration forces. It can be seen that in the maximum bolt force case (top diagram) the threads have an additional tensile force of 5,400 pounds acting on them. In the minimum bolt thrust case (bottom diagram) the threads have a compression force of 800 pounds acting on them as a result of the cartridge case forces. Recall that the tension force (5,400 pounds) derived from the axial preload stabilizes the joint and the compression force (800 pounds) destabilizes the joint. However, this only occurs during the time the chamber is pressurized and the direction of the force will reverse when the chamber pressure drops. What this means is that the large stabilizing force of 5,400 pounds can reverse sometime after the bullet leaves the muzzle and become a 5,400 pound destabilizing force. We don't know when the joint actually moves. It is most likely happening shortly after bullet exit, because that is when the loads that reduce the axial preload appear to be the largest. This is also borne out by the muzzle scope tests where the barrel is observed to move after a shot that was in the group, but the next shot would be wild.

There is another dynamic load that acts in a similar manner to the bolt thrust, that is due to chamber radial expansion from the chamber pressure. One can calculate that this radial expansion produces an axial tension load of somewhere between 8,250 and 11,512 pounds depending on the degree of constraint provided by the forward receiver ring. The correct value is probably somewhere in between, so let's assume a value of 10,000 pounds.

In addition, there is an impact force on the joint when the recoil lug impacts the stock in a standard sporter. You may recall that this was measured back in Chapter 4 and found to be 1,500 pounds. This is probably a lesser effect on bench rest actions that use pilar or glue-in bedding. Also, the recoil force is less on the 6PPC than the 270 Winchester. However, there must be some "recoil lug" force still acting on bench rest rifles because the recoil force must be transmitted from the action to the stock.

If we sum up all these destabilizing loads from different sources for the case of a cold and a hot barrel we get:

Cold Barrel	
Recoil lug	1,500
Bolt thrust	5,400
Chamber radial expansion	10,000
Total	16,400

Hot Barrel	
Recoil lug	1,500
Bolt thrust	5,400
Chamber radial expansion	10,000
Differential temperature expansion	7,000
Total	23,900

What all this comes down to is that the joint with an axial preload of 20,000 pounds has marginal stability under ordinary conditions, and is unstable when hot after firing two or three 5-shot groups without cooling. This roughly corresponds with our experience shown in Table 5, so I feel certain that it is a real effect. What is needed is an axial preload in excess of 24,000 pounds to assure that the joint cannot move under these extreme conditions of heating, shock, and vibration.

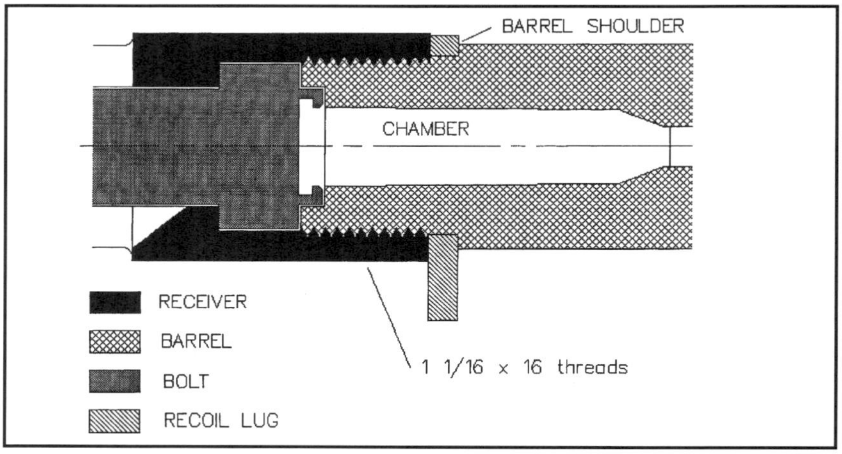

Figure 6-4 - Drawing of the standard Remington 721 barrel joint design with National Form 60° V threads and recoil lug. It is equivalent to a standard bolt and nut joint, where the recoil lug serves as a washer.

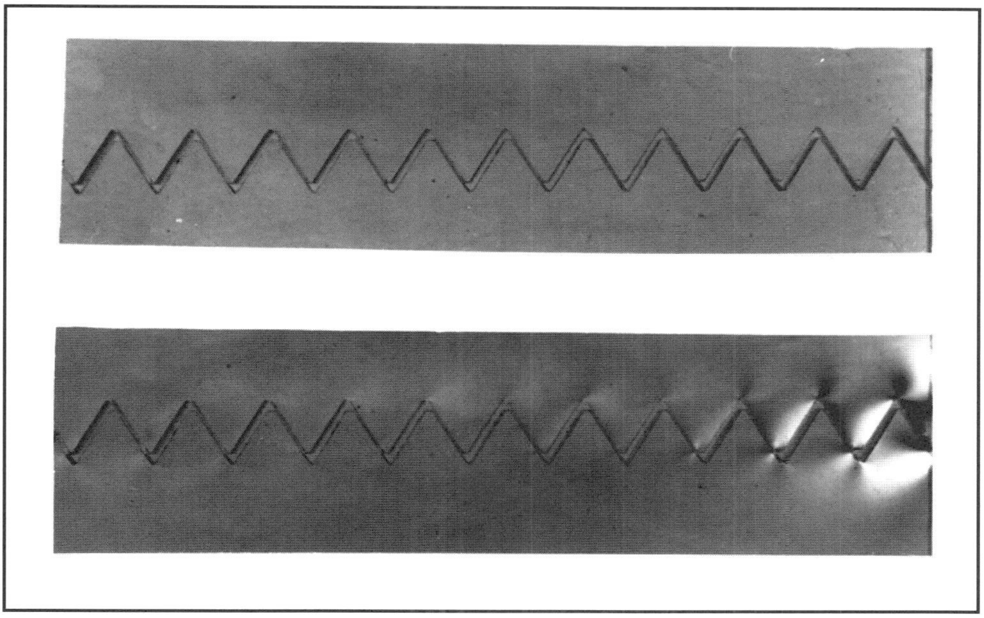

Figure 6-5 - Photograph of a plastic model of the barrel joint threads (standard V thread), using polarized light, which demonstrates how the load on the threads is concentrated on the first few threads near the front of the receiver (bottom right). Top photo shows the threads in the unloaded condition.

Joint Redesign

We have already redesigned the recoil lug to take the higher loads by making it out of a stronger steel (4140) which can be heat treated to a higher yield strength (130,000 psi). The commercial lug appears to be stamped out of a mild steel, which can't be heat treated much above 60,000 psi. I also silver brazed the recoil lug to the front of the receiver (Figure 6-4). This was done primarily as a convenience to keep the lug from rotating during all the barrel changing that had to be done. However, it may have some effect on joint stability, and I believe this should be done during manufacture. The barrel shoulder will take a load of about 32,000 pounds before yielding, so we will try to come up with an improved thread design that will stand an axial preload of this level.

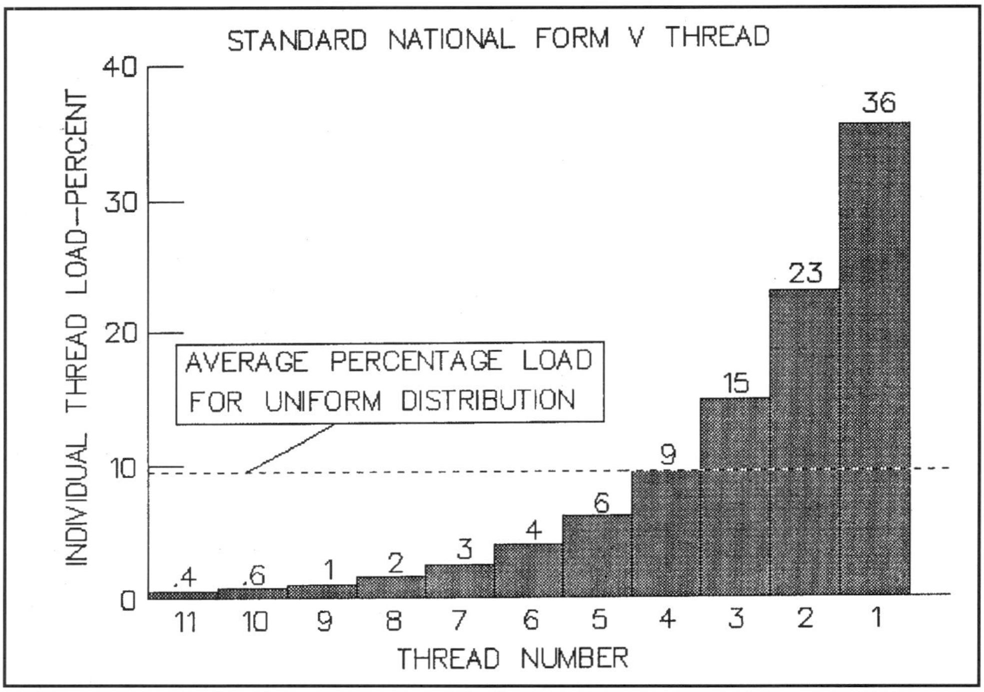

Figure 6-6 - Theoretical calculation of the individual thread load as a percentage of the total preload. The front of the receiver is on the right of the figure. Only the first few threads carry a significant load.

Standard National Form V Thread

The standard National Form V thread used on this joint is the same thread that is used on most bolts and nuts. While this is a well known situation (References 18,19) in mechanical engineering circles, it will likely surprise the reader to find out that the individual threads are not equally loaded. It turns out that a large percentage of the total axial preload is carried by the first few threads next to the front of the receiver. This is demonstrated in Figure 6-5 which shows the result of a photoelastic test. The threads are machined into two plastic sheets that are 1/8 inch thick. The threads are four times the size of normal 16 thread per inch threads found on the Remington action (i.e., 4 threads/inch). The two pieces are held in a fixture while a force is applied to the right side of the top piece acting to the left, and an opposing force acting to the right is applied to the bottom piece. The test is run using polarized light, which is distorted by stresses in the plastic caused by the load. The plastic model is back lighted and is sandwiched between two pieces of polarized film. The two films are rotated relative to each other until the maximum effect is obtained and then photographed. Tension stresses show up as a dark area and compression stresses show up as a light area. Both the unloaded case (top) and loaded case (bottom) are shown in the figure, and it can be seen that most of the load is taken up by the first few threads near the front of the receiver. Obviously, the threads toward the chamber end of the barrel on the left side of the figure are essentially unloaded. The load on each thread in terms of percentage of the total load was calculated using the method of Bluhm and Flanagan (Reference 19) and is shown in Figure 6-6. You can see that the first thread nearest the front of the receiver carries about 36% of the total load, while the last several threads carry only a small percentage of the total load. This explains why the joint is really not “frozen” at all, and is relatively free to move, because only the first three or four threads carry a

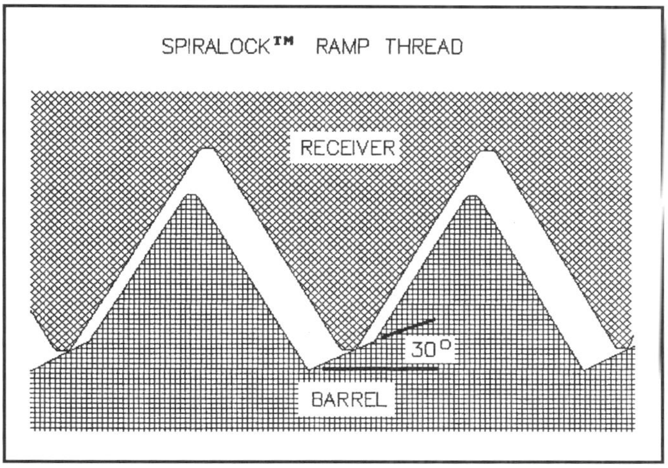

Figure 6-7 - Sketch of the Spiralock™ ramp-thread. The width of the 30° ramp is 0.025 inch compared to the pitch or width of the thread of 0.0625 inch. Only the peaks of the receiver threads are loaded. The threads are shown in the loaded condition. The front of the receiver is to the right.

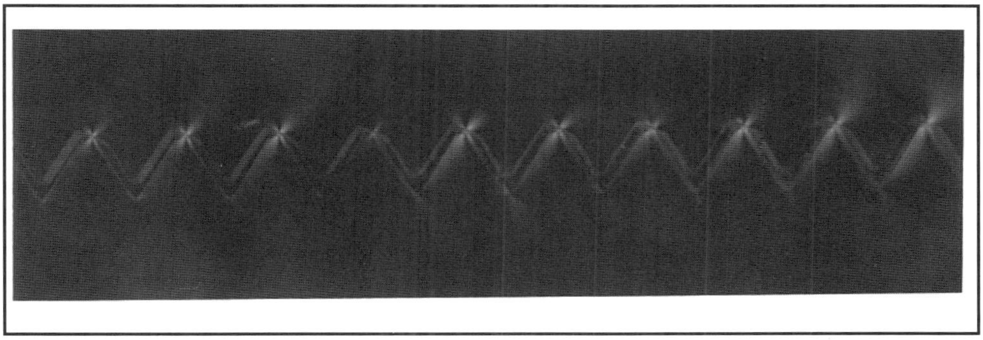

Figure 6-8 - Photograph of plastic model of the barrel joint showing how the ramp-thread has more evenly loaded threads, which allows the preload on the joint to be increased. Front of the receiver is on the right.

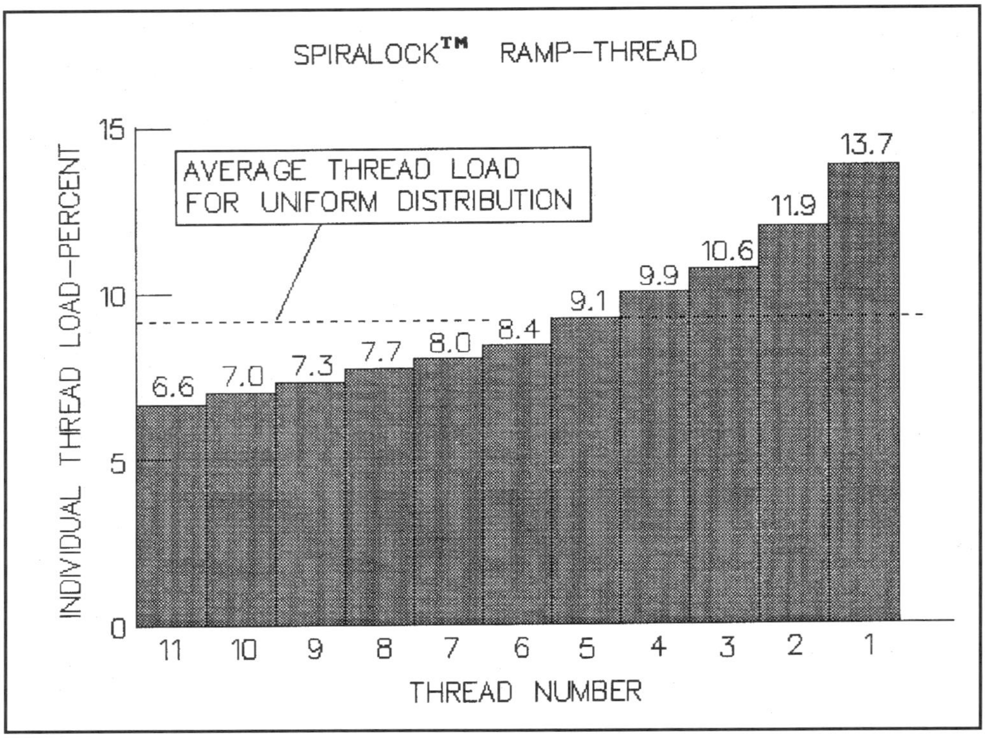

Figure 6-9 - Calculated individual thread load for the Spiralock™ ramp-thread in terms of percentage of total preload. The load distribution is much more evenly distributed than was the case with the standard thread (Figure 6-6).

significant axial load. If the threads at the rear end of the barrel were loaded to roughly the same extent, then the barrel joint would not be as likely to rotate about a lateral axis and would be more likely to stay aligned with the receiver. The first thread carries a load of 7,200 pounds with an axial preload of 20,000 pounds. Since we found out that this is the axial preload level where the threads first start yielding, we could theoretically load this joint up to 79,200 pounds if the load distribution on all eleven threads was perfectly even! This is an impractical load level because the barrel shoulder would fail. Fortunately, there are at least three ways of improving the load distribution on the threads.

Ramp-Thread

A sketch of the ramp-thread, which is a patented thread design called Spiralock™ by Detroit Tool Industries, is shown in Figure 6-7. A photoelastic experiment performed by the author (Figure 6-8) shows that the threads are more evenly loaded than the previously tested standard V thread. Data supplied by Detroit Tool Industries were used to calculate the load distribution shown in Figure 6-9, where it can be seen that the load distribution is much improved. For instance, the first thread only carries 13.7% of the load compared to 36% for the standard thread. Also, the threads toward the chamber end of the barrel carry a much larger percentage of the average load. Well, I machined a new barrel (chrome moly 4140) with the ramp-thread and installed it with a measured axial preload of 32,000, using Teflon tape as a lubricant and a torque of about 200 foot-pounds. After firing 50 rounds the axial preload had dropped to 29,000, and there was no visual observable evidence of metal yielding (i.e., permanent deformation). This 10% loss in axial preload is probably due to small local yielding at stress concentrations, and is to be expected with a new threaded joint. The ramp thread joint was reassembled with an axial preload of 31,000 pounds and test fired again at a rapid rate to heat the barrel. The wild flyers that had been present at elevated temperatures with the standard V thread were eliminated, indicating that this joint design is successful. Since the ramp thread can be cut on the barrel with a die or with an ordinary threading tool modified to have the ramp shape, this design modification is practical in production. Barrel replacement by gunsmiths is also practical. Tolerances are not any more critical than on the standard V thread. This approach was suggested by my son, who is a mechanical engineer. Since some of the patents on this thread are still in force, Detroit Tool Industries may require a patent release for large scale production. However, the President (Mr. Ed Palm) of Detroit Tool Industries (1-800-521-2688) assured me that they would not object to custom gunsmiths using this thread. There is another approach that is probably not patentable, and that is a variable depth thread similar to a pipe thread with much less taper.

Variable Depth Thread

The reason that the standard V thread doesn't have a uniform load distribution, is that when the joint is tightened the barrel stretches and the receiver compresses in the longitudinal direction. This causes the threads toward the chamber end of the barrel to be unloaded. While this only amounts to 0.5 mil (1 mil total difference between receiver and barrel), it is enough to almost completely unload the last thread. What we need is a thread that starts being loaded on the end of the barrel, and as the joint is tightened, transfers some load to the threads near the front end of the receiver. A friend of mine (John Weydert) who is a mechanical engineer, suggested machining the threads in the barrel at the front of the receiver a little deeper than those on the chamber end of the barrel, using a linear taper. I made calculations that predicted that a variation in thread depth of 2.5 mils/inch would result in uniform loading for an axial preload of 30,000 pounds. The amount of taper depends on the cross section area of the barrel tenon. This thread is easily machined on a lathe using a taper attachment. However it would be easier and better to cut the tapered thread in the receiver during production using a tapered tap. A photoelastic test of the tapered depth thread (Figure 6-10) showed that the load on the individual threads was fairly uniform. You can see in the photograph that the clearance between the threads on the right side of the figure (receiver front end) is greater than on the left, which corresponds to the end of the barrel. The variable depth is much exaggerated in this figure, because plastic is much more elastic than steel. There is no point in a theoretical calculation, because it will simply predict a constant load of 9.1% of the total

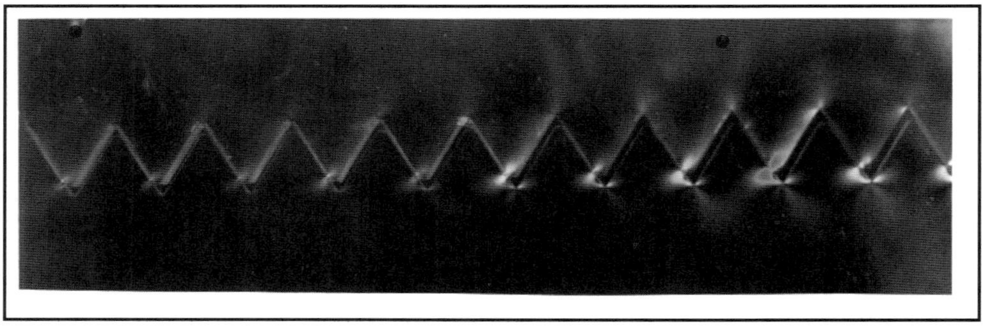

Figure 6-10 - Photograph of plastic model of the barrel joint showing how the tapered-depth thread has more evenly loaded threads, which allows the preload on the joint to be increased. Notice the greater depth of the threads at the front of the receiver on the right side.

load per thread. On the strength of this photoelastic test, I decided to machine another barrel and install it on another used Remington 721 action, that I had purchased, for testing. The barrel with the variable depth thread was installed with an axial preload of 27,200 pounds. After firing, the axial preload dropped to 24,500 pounds, which seemed to be the maximum axial preload that this thread will sustain. This is not as good a design as the ramp thread, but it is better than the regular V thread. After this test I decided to stay with the ramp thread, because I know it works.

Ramp Thread Accuracy Test

The accuracy was tested and the average group size was essentially the same (0.884 inches) as that presented in Table 4 in a previous chapter (Chapter 4). There were no large flyers at high temperatures like there were before the new ramp-thread joint was installed. In the previous accuracy test the gun was cooled after every other group (i.e. 10 shots), where this time I fired four 5-shot groups rapidly before cooling and cleaning.

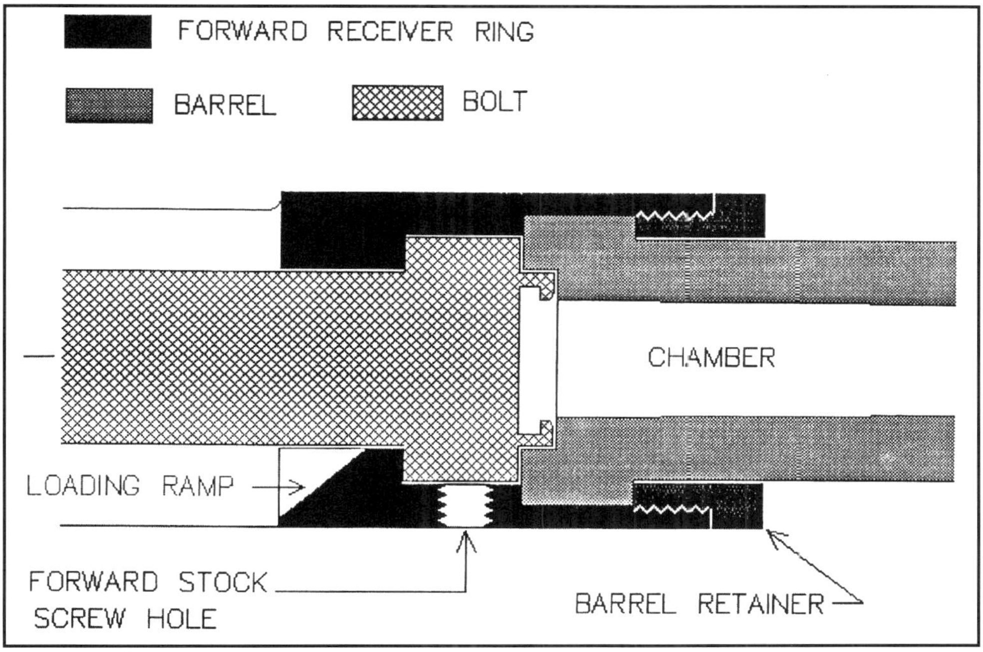

Figure 6-11 - Cross-section drawing of a barrel joint design that does not depend on a large preload for stability and is not affected by temperature gradients.

Figure 6-12 - Photograph of an action showing a disassembled barrel joint design that is immune to temperature and other barrel joint problems. This design is also depicted in Figure 6-11.

Complete Barrel Joint Redesign

A sketch of an improved barrel joint design is shown in Figure 6-11, and a photograph of a receiver and barrel incorporating the improved design is shown in Figure 6-12. The collar on the barrel is clamped between a shoulder in the receiver and the threaded retainer ring. Consequently, thermal expansion of the barrel or collar in the axial direction simply increases the force holding the joint together instead of relieving the force as it does on the normal barrel joint. In other words, when the barrel heats up as a result of rapid firing or the barrel stretches as a result of the action of the cartridge case, the joint just gets tighter. This design is a much more reliable approach because it doesn't depend to any great extent on the magnitude of the axial preload. As long as the retaining ring is reasonably tight the barrel is locked in place. The fact that the design works was proven with the hardware shown in Figure 6-12. This particular receiver and barrel could be fired with the scope on either the receiver or on the heavy barrel. The barrel measured 1.2 inches in diameter at the chamber end and 0.9 inches at the muzzle. The average 5-shot group measured 0.65 inches and there was no difference in group size between the two scope locations. This is strong evidence that this barrel joint design is stable and does not allow motion between the barrel and the receiver between shots.

There is one other way of preventing barrel joint motion and that is to make the receiver and barrel out of one piece of steel. The most accurate commercial rifle that I ever had (Savage Model 23D in 22 Hornet) was made this way. Unfortunately, this is impractical because you can't replace the barrel.

Summary

It was demonstrated that the standard V thread barrel joint was moving. Also, the barrel motion was particularly severe under the high temperature conditions obtained by rapidly firing 15 rounds, which caused large flyers. Theoretical calculations as well as experimental measurements indicated that the barrel joint could move when the barrel was hot if the axial preload was less than 24,000 pounds. This problem was corrected by changing to a ramp-thread that allowed increasing the sustained joint axial preload from 20,000 pounds to around 30,000 pounds. The increased axial preload obtained with the ramp-thread prevented any barrel joint motion.

The question will arise as to whether all threaded barrel joints move in bolt action rifles. Obviously, no one knows, but I think it is likely that all bolt action rifles with threaded barrel joints probably do move to some extent, although those actions with integral recoil lugs and standard bedding probably are less effected. O'Connor bedding, which was discussed back in Chapter 4, may stabilize the joint to some extent. O'Connor bedding applies a preload to the joint which likely stabilizes it. Most engineers know that it is very difficult to make a rigid threaded joint, particularly under the temperature, shock and vibration conditions present in a rifle.

Bench rest rifles have heavy barrels which conduct the heat away from the barrel joint reducing the effect of temperature. Also the heavy barrel helps reduce the load on the joint. They often have the receiver bonded to the stock reducing the effect of recoil on the joint. These features coupled with the fact that smaller calibers are usually used, and the barrels are cleaned frequently allowing the temperature differential to equalize, reduce the probability of barrel joint motion on bench rest rifles. However, barrel joint motion is a fact of life and it can be present to some degree in any rifle. I am very suspicious of the short handled action wrenches used by bench rest shooters to install or change barrels. These short wrenches make it impossible to apply sufficient torque to obtain a satisfactory axial preload with ordinary lubrication. Increasing the axial preload by using Teflon tape as a lubricant will reduce the tendency of the barrel joint to move. Using Teflon tape and the same applied torque would more than double the axial preload.

CHAPTER 7
MUZZLE BLAST

More than thirty years ago a friend of mine (Ed Cave) and I were camped in a mountain meadow covered with tall green grass. Green grass is pretty unusual in this part of the world (New Mexico). I decided to shoot at a target on a distant mountain side, so I sat down and fired several rounds over the grass. My friend was standing behind me, and when I had finished shooting he said, "Something funny is going on—sometimes I don't see the muzzle blast on the grass and at other times it appears off to the left or to the right and sometimes right in front of you." Well normally I would have answered "Uh huh" and gone on shooting, but I knew this guy was an accurate observer. So, I asked him to fire a few rounds so that I could watch, and sure enough he was right. It was pretty clear that the direction of the muzzle blast varied a lot from shot to shot. Well, that experience has bothered me for years, so I decided to find out just what the heck was happening. The first thing to do was to repeat the "grass" experiment in a more professional manner, and try to get some presentable data. Since I am an old aerodynamicist, I decided to use an "Old Aerodynamicist's Trick." Back in the good old days, when we couldn't tell what was going on in an airflow problem, we used to attach things called tufts to a wing or some other shape, which allowed us to tell which way the wind was blowing. These tufts were made of two or three inch lengths of wool yarn. They would follow the direction of local air flow, and were very helpful in diagnosing aerodynamic problems. Sometimes, hundreds of tufts were required. These days computers are used

to simulate the flow, and usually do a better job. I decided to apply the old tuft technology to the muzzle blast problem, because it is simple, cheap, and you can get photographic data.

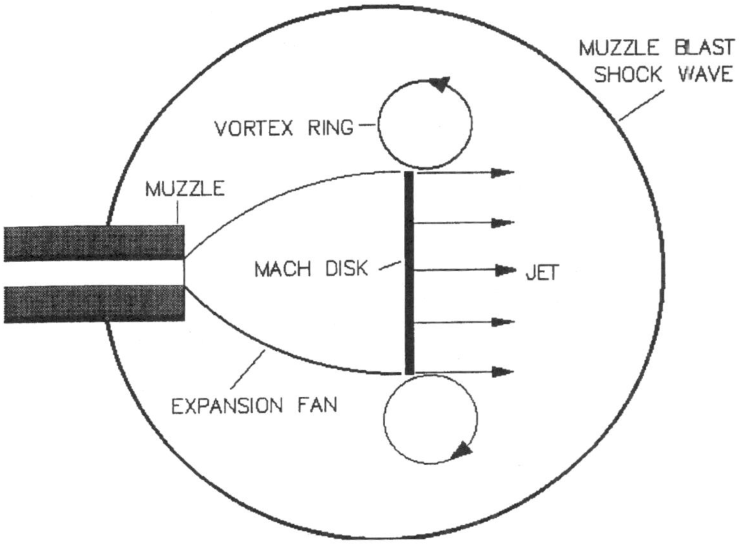

Figure 7-1 - Diagram of the muzzle blast flow field showing essential features.

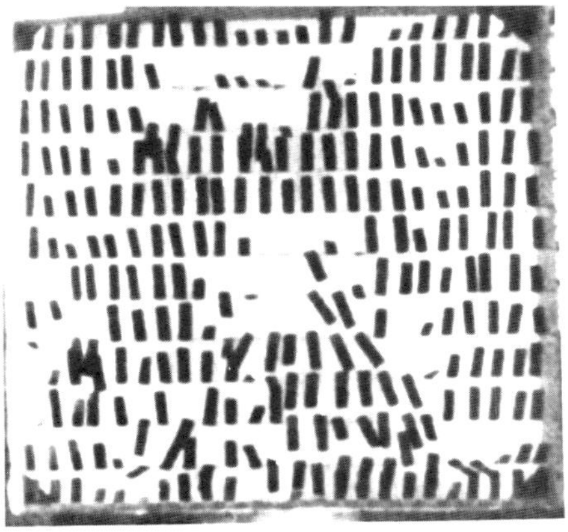

Figure 7-2 - Photograph of the tuft screen showing centrally located muzzle blast. Screen is 4 feet square and was placed 18 feet from the muzzle. The circular pattern around the jet impact in the center results from an annular vortex ring, similar to a smoke ring.

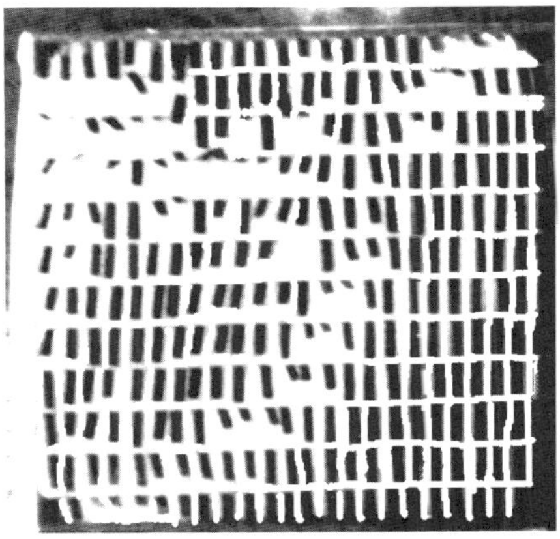

Figure 7-3 - Photograph of tuft screen where the muzzle jet was deflected up and to the left. A segment of the vortex ring extends from lower left to upper right.

Tuft Screen

A four foot square frame was built of 1X6 inch wood with 2X4 inch screen wire attached to the front of the frame. A white sheet was attached to the rear of the frame to enhance the photographic contrast. The tufts were made of 1 inch wide black paper. They were attached to the wire grid by bending the paper over the wire and fastening with Scotch tape. A bullet was fired through the center of the screen, which was placed about 18 feet in front of the muzzle, and the screen was photographed with a Polaroid camera. The idea was that the tufts would be displaced to the rear by the muzzle jet and its associated flow field. We should then be able to determine the location of the center of the muzzle blast flow field from the photographs. Figure 7-1 shows a simplified sketch of the muzzle blast flow field. This flow region is called the transitional ballistics region between internal and external ballistics. When the hot, high pressure gas exits the muzzle it expands and the flow becomes supersonic. As the gas expands it slows down to sonic velocity (Mach number one) and generates a thick shock wave called a Mach disk. The jet velocity at the Mach disk is roughly 4000 fps because the temperature is high (~6000°F) and the speed of sound is high. This high speed flow generates a vortex ring similar to the smoke rings that smokers sometimes make. The muzzle blast shock wave, which produces the loud noise when you fire a rifle, continues to expand and eventually becomes a weak sound wave. The jet and the vortex ring continue to travel for some distance (at least 20 feet) and these are what we see hitting the tuft screen. Figure 7-2 shows a photograph of the tuft screen where the muzzle jet has hit the center of the screen. If you look closely you can see a rough circular pattern around the center of the screen. This circular pattern is caused by a vortex ring that forms around the jet in the center. Figure 7-3 shows a case where the muzzle jet has hit the upper left corner of the screen, and a portion of the vortex ring can be clearly seen extending from the lower left to the upper right. This indicates a jet angle of about six degrees from the bore axis. The next figure (Figure 7-4) shows a case where both the jet and the vortex ring completely missed the screen, which requires a muzzle jet angle of at least 12 degrees. The white square in the center of the screen was the aiming point where there are no tufts. Four different brands of 270 bullets were used, and all but one had large jet deflection angles. Missing the screen entirely was the most typical result, and it indicated that I should have used a larger screen. However, this size screen involved making 284 tufts, which was a lot of work. It probably

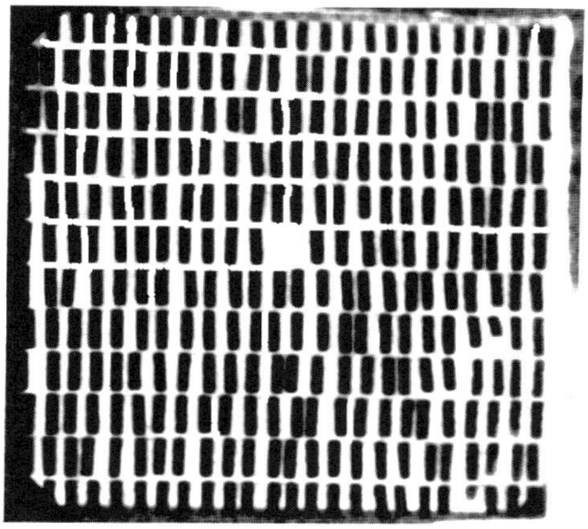

Figure 7-4 - Photograph of the tuft screen showing a case where the muzzle blast missed the screen entirely.

would have been better to take high speed motion pictures of each shot, and make a print of the frame that occurred at the optimum time. However, I think that the results are good enough to show that the muzzle jet was not centered with some bullets. This verifies what my friend and I observed while firing over grass a long time ago. I should point out that the tufts do not respond to the shock waves generated at the muzzle, but do respond to the jet issuing from the muzzle and the vortex ring, which slow down rapidly. The only reason for doing this experiment was to show the reader that the muzzle blast could be asymmetric with respect to the bore axis at some distance from the muzzle, because the data are useless as far as predicting the effect on bullet dispersion. But first, what could be causing this problem?

Bullet In-bore Cant

The only logical way for the muzzle blast shock wave pattern to be asymmetrical with respect to the bore axis is for the base of the bullet to be canted when it exits the bore. The most likely way for the base to be canted is for the bullet to be canted while it is in the bore. I had observed unequal circumferential rifling engraving on recovered cannon shells, which confirmed the fact that cannon projectiles can cant in the bore. However, I was reluctant to believe that a standard tangent ogive-cylinder (see Figure 7-5) rifle bullet shape could cant or tip in the bore. So, I decided to inspect some of the 270 bullets to see if there was some defect in the shape of the bullets that would allow them to cant while in the bore.

A number of bullets from different manufacturers were examined. Practically all of them had a conical or ogive shaped afterbody instead of a true cylindrical shape. The afterbody is the rear portion of the bullet that extends forward from the base to where the ogive starts tapering. A conical or

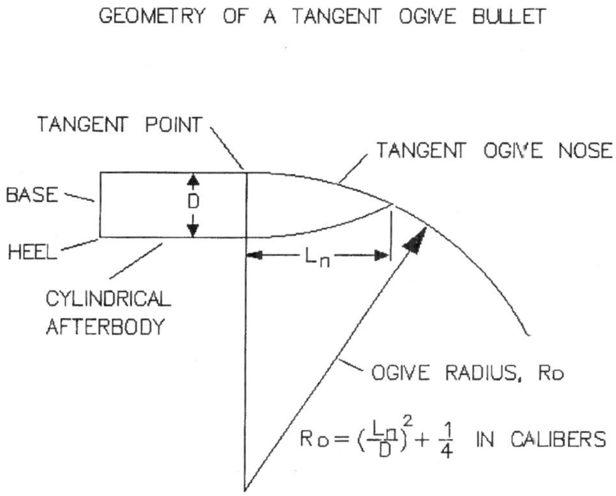

Figure 7-5 - Drawing showing the geometry of a tangent ogive bullet shape.

tapered afterbody will certainly allow a bullet to tip when it travels down the bore. On many bullets the ogive shape continues aft to the base and there is no cylindrical afterbody. My guess is that the manufacturers do this to make it easier to eject the bullets from the forming die. Having made a few bullets, I can guarantee you that ejecting a bullet from a die can be difficult, and a slight taper on the afterbody would greatly alleviate the problem. Figure 7- 6 is a photograph of a 90 grain Hollow Point 270 bullet held between the parallel jaws of a dial gage micrometer. It can be seen that the afterbody starts tapering to a smaller diameter immediately in front of the base. Figure 7-7 shows a photograph of a Remington 130 grain 270 Bronze Point that has a true cylindrical afterbody that is about 1.2 calibers in length. Unfortunately, this bullet has a canelure groove, which reduces the length of the cylindrical afterbody. It also reduces the ballistic coefficient, and can cause bullet imbalance. It should be pointed out that 90 and 100 grain 270 bullets can't have much more than a 1.5 caliber cylindrical afterbody, otherwise the nose would be too blunt. Well, it is clear that if we fired the bullet shown in Figure 7-6, it could tip in the bore if it were fired from a smoothbore. However, the sides of the bullet are contacted by the rifling lands about 1.5 calibers ahead of the base despite the tapered afterbody, which should help to stabilize the bullet and keep it from tipping. So, it seemed possible that the bullet might not actually cant in the bore, in spite of the fact that the afterbody is tapered. The only way that I could be sure that bullet cant was the cause of the asymmetrical muzzle blast problem, was to recover some bullets and examine them.

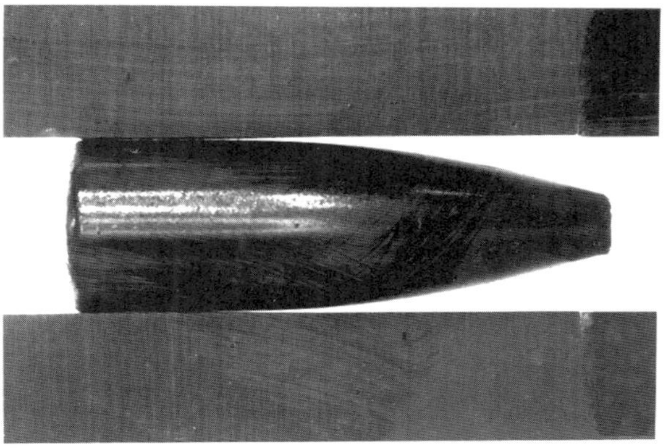

Figure 7-6 - Photograph of a 90 grain 270 HP bullet held between the parallel jaws of a micrometer, showing the tapered afterbody.

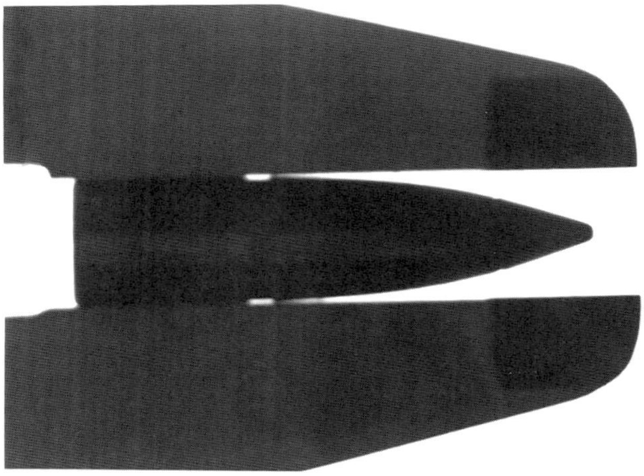

Figure 7-7 - Photograph of a Remington 130 grain 270 Bronze Point bullet held between the parallel jaws of a micrometer, showing the parallel sides of the cylindrical afterbody.

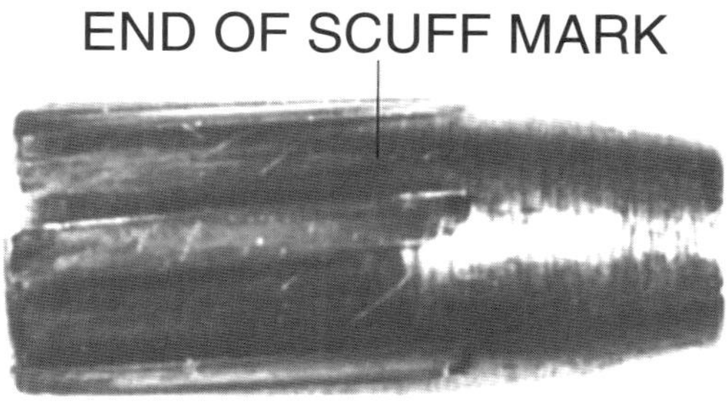

Figure 7-8 - Photograph of a recovered 90 grain 270 HP bullet showing triangular shaped scuff marks on the side wall of the bullet between the rifling marks, which can be measured to determine bullet cant.

Bullet Recovery

Recovering a bullet undamaged at high speed is tough to do, so I decided to cut the nose off (Figure 7-8). This increases the drag coefficient by a factor of 2.5 and helps slow it down. The calculated velocity at 300 yards was 1400 fps. That was where I placed a 1.5 foot square by 4 foot long plywood box filled with pine sawdust. The 90 grain 270 HP bullets, which weighed 86 grains after the nose was trimmed, traveled about 3.5 feet through the sawdust and were recovered in good shape. This was farther than I had expected (2 feet), probably because the sawdust density (13 pounds per cubic foot) was less than I had expected (20 pounds per cubic foot). The 130 grain bullets went through the whole thing and were not recovered. If you want to recover the heavier bullets in less than four feet, drill out some of the lead in the nose to reduce the weight to less than 90 grains, use a longer box, or recover them at longer range. It may be a problem to adjust the sights at longer range to correct for the bullet drop. In this case it was 31 inches, which was precalculated and turned out to be correct. This weight can be scaled by the sectional density for other calibers.

The angle of bullet cant can be determined from the recovered bullets by measuring the difference in length of the rifling marks from side to side. A simple equation can be derived that can be used to approximate the angle of cant.

cant angle   28.65*L/Ro degrees

where

L = differential length of rifling marks (inches)

Ro = radius of the tangent ogive nose (1.2 to 1.5 inches)

The recovered 90 grain bullets were measured and L was found to range between 10 and 20 mils. This meant that the cant angle ranged between 0.24 and 0.48 degrees for an Ro of 1.2 inches. This agrees well with the measured afterbody angle that was obtained from Figure 7-6. So the tapered afterbody can allow the bullet to cant in the bore by as much as 0.5 degrees. I found that the best way of measuring L was to measure the length of the scuff marks where the side of the bullet contacted the groove between the rifling lands. This scuff mark, which is roughly triangular in shape can be seen in

Figure 7-8. It is difficult to photograph and is much more obvious to the eye. The triangular shape is due to the taper in the afterbody of the bullet.

It is possible that this method of measuring bullet cant is in error by as much as a factor of two. The bullet may start out in the throat in the uncanted condition, causing the rifling marks to be even in length at first. The bullet may then cant further down the barrel, when maximum pressure is reached. This will lengthen the rifling marks on one side of the bullet, but cannot shorten the rifling marks on the other side. The same thing will happen to the scuff marks caused by the rifling grooves. Consequently, you can't be sure that you have measured the maximum cant angle from recovered bullets, but I can't think of any better way to do it.

We now know that bullets can cant in the bore, which could be caused by a tapered afterbody. But, how much dispersion could this cause?

Bullet Base Cant Dispersion

In order to simulate the effect of bullet cant on dispersion I relied on an "Olde Engineers Trick". Simply stated, it means that if you have a small error, that is difficult to measure, exaggerate the error so that it can be measured. For this approach to work you have to be careful not to change anything else. This can be done by cutting off the base of the bullet at a two degree angle

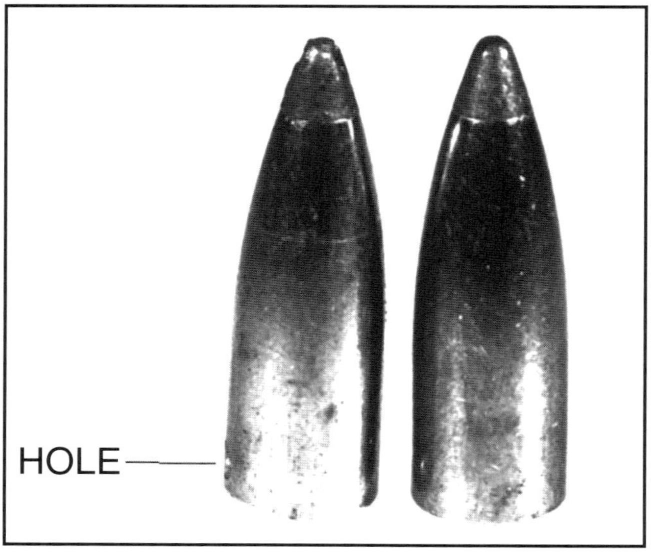

Figure 7-9 - Bullet on left with base milled off at a 2 degree angle compared with normal bullet on right. Small hole in the left side of the base of the modified bullet compensates for mass asymmetry caused by the slanted base. The bases of both bullets are resting on plate glass. Note that the modified bullet on the left leans toward the normal bullet on the right.

and then drilling a hole on the long side of the bullet to compensate for the shift in center of gravity (CG). This modification can be seen in Figure 7-9. The two degree base angle is about 4 times the maximum amount of bullet cant that I measured. It requires removing about 10 mils from the jacket base on the short side of the bullet. Since the jacket is only about 25 mils thick at the base, a 2 degree angle is about all the bullet can tolerate. A larger angle would result in thinning the base jacket so much that there would be a real risk of blowing the lead core through the jacket during firing. This could leave the jacket lodged in the bore, which is a very dangerous condition. It is also important to use a bullet with relatively sharp corners on the base, which minimizes the amount of material to be removed from the jacket at the base of the bullet. For this reason, 100 grain 270 soft point bullets were used in the experiment. The test routine is to fire four 3-shot groups with the base asymmetry pointed up, right, down, and to the left at muzzle exit. If you perform this test, it is necessary to remove the extractor and ejector from the bolt, and to use cases with at least 5 mils of headspace. Otherwise, the bolt will rotate the cartridge and mess up the experiment. The cases have to be ejected with a cleaning rod, so be sure to remove it before firing the next shot! It is a good idea to look down the bore just to be sure it is clear before inserting the next live round. The only way that I have found to index the cases in roll angle is to put a mark on the head of the case. The mark is rotated until it is properly indexed, and then the case is pushed all the way into the chamber with a finger. It isn't easy but it can be done. This gives us four distinct groups spaced roughly 90 degrees apart, which tells us a lot about just how this base cant asymmetry works. The results of this test are shown in Figure 7-10. The mathematical centers of the groups are shown by squares, and the direction of the short side of the bullet is indicated. The bullet should be deflected in the direction of the short side of the bullet if there were no gyroscopic action. You can see that the groups are all rotated clockwise somewhere between 30 and 60 degrees. This means that the bullet was disturbed over the first 6 or more calibers of its travel after leaving the muzzle, which is to be expected. Schmidt and several coworkers at the US Army Ballistic Research Laboratory (Refs. 4 and 5) have shown that the bullet is influenced by the muzzle blast for roughly 15 to 20 calibers. The stronger pressure, which has the biggest effect, is present much closer to the muzzle. Just why the group rotation is not more consistent is not known. However, there are other effects including the error in indexing the cartridges, bullet imbalance, statistical error in 3-shot groups, and other effects that could

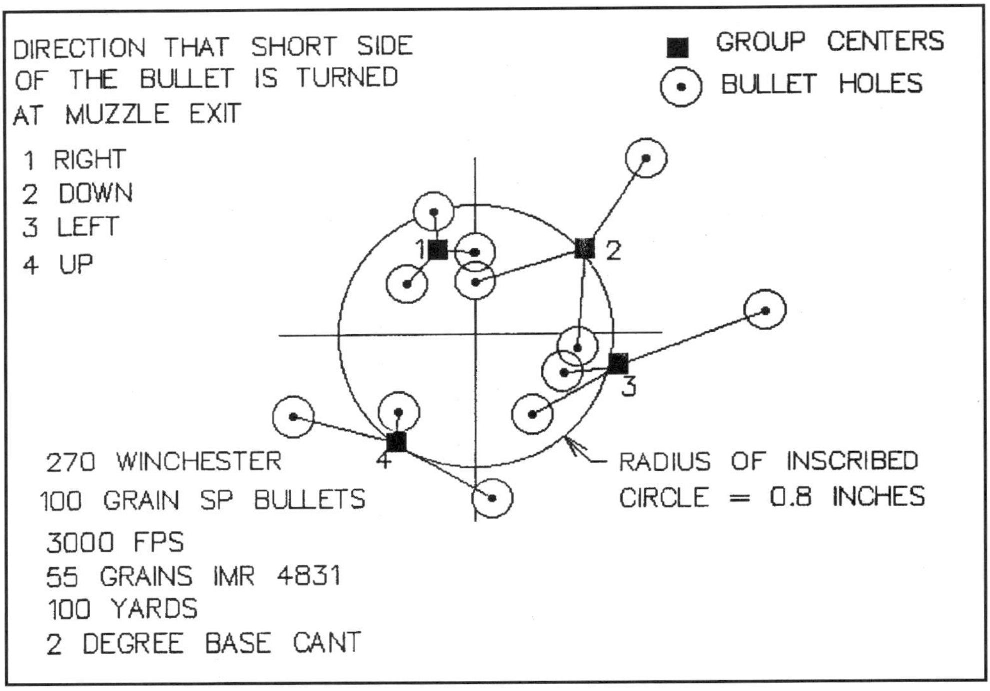

Figure 7-10 - Plot of the target showing the results of firing four 3-shot 270-groups with the slanted bases of the bullets indexed at 90 degree increments in roll angle. Square symbols indicate the center of each group. The radius of the inscribed circle (radius of dispersion) is 0.8 inches.

easily explain the variation in rotation angle of the groups. The interesting finding is that a radius of dispersion of 0.8 inches resulted for a base cant angle of 2 degrees. This means that the maximum 0.5 degree bullet cant angle that we measured would result in group sizes approaching 0.4 inches if all other sources of error were eliminated. Even though this is a relatively crude experiment, I believe it shows that bullet cant causes dispersion. Therefore, bullet cant is a significant source of dispersion, even though it is smaller than some of the other error sources.

After working on this problem for over a year I was able to run the same test in a rail gun chambered for the 6mm BR. The test was run in a tunnel range which eliminated wind effects. This equipment was mentioned in Chapter 4 and described in detail in Appendix E and F. The results of the 6BR test with the bases of the bullets milled off at a 2 degree angle are shown in Figure 7-11. You can see that the results are similar to the results for the 270, except that the radius of dispersion is smaller (0.64 compared to 0.80).

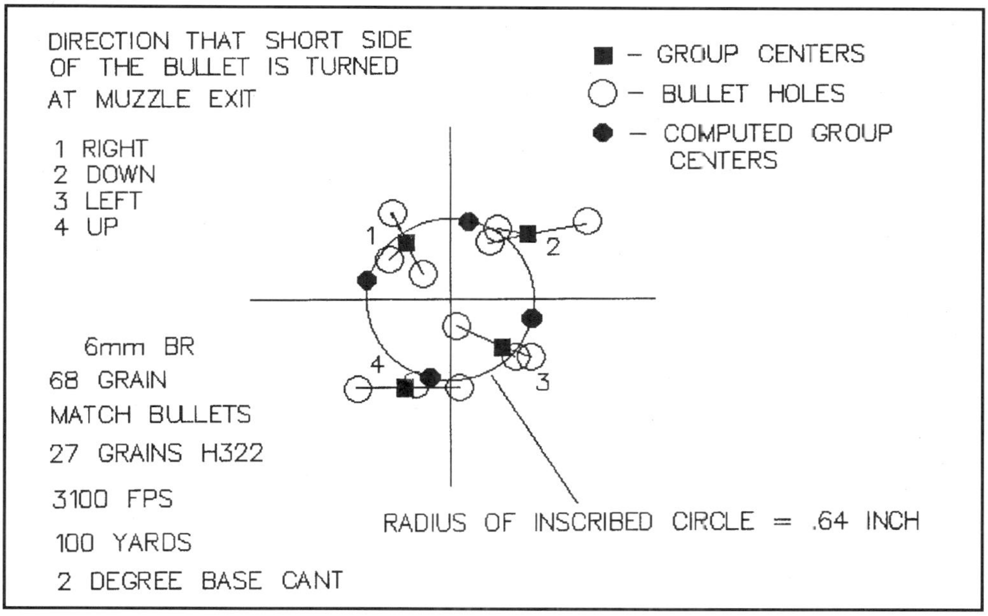

Figure 7-11 - Plot of the target showing four 6mm BR groups fired with the bases cut off at a 2 degree angle and indexed at 90 degree intervals in roll angle. Radius of dispersion is 0.64 inches and is predicted by theory (solid circle symbols).

The fact that the radius of dispersion was smaller indicated it might have something to do with the lower muzzle blast pressure of the 6BR. We will measure the effect of different powders on muzzle blast pressure later in this chapter.

Canted Bullet Test

Since the canted base test was only a test to see if there was a muzzle blast effect I decided to try testing the real case where the whole bullet is canted in the bore. The 6BR was tested by canting the bullets 0.215 degrees in the case neck and indexing them in roll just as before. The result was a group that looked like a four leaf clover and is shown in Figure 7-12. The radius of dispersion was 0.196 inches and is shown by the inscribed circle. The theoretical value obtained from a trajectory simulation computer code was a little larger and was 0.243 inches. The 70 degrees clockwise rotation of the groups is predicted by the theory. There are at least two explanations why the experimental value was slightly smaller than the theoretical value. The bullet may

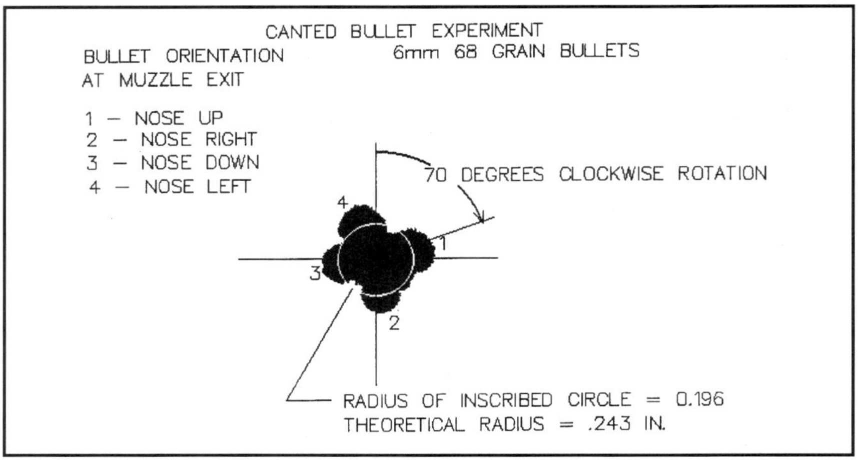

Figure 7-12 - Computer scan of the target where four groups of 4 bullets each that were canted 0.215 degrees in the case neck. The bullets were indexed in 90 degree intervals in roll. The radius of dispersion of 0.196 inches is superimposed. The trajectory simulation predicted 0.243 inches for radius of dispersion.

have straightened out slightly upon entering the throat or the Center of Gravity (CG) may have been slightly away from the side of the bullet most in contact with the bore. The bullet would only have to be pushed sideways about 0.05 mils to cause enough CG offset to explain the difference between the theory and experiment. Either one or both of these things are physically reasonable and may have happened. The important thing to realize is that only a 0.2 degree canted bullet angle resulted in a radius of dispersion of about 0.2 inches. This is a much larger effect than I would have expected from the canted base tests. You see, the radius of dispersion for the canted bullet tests was about 1 inch per degree of bullet cant while the radius of dispersion was only about 0.32 inches per degree of base cant. I would have expected the sensitivity to be about the same, but it wasn't. One other thing that I noticed is that you have to back the canted bullets off the lands to obtain these results, otherwise the bullet misalignment will be reduced. The first time I tried the canted bullet experiment the bullets were in contact with the lands and the radius of dispersion was only 0.15 inches. The test shown in Figure 7-12 had the bullets backed off the lands 30 mils. Most bench rest shooters push their bullets forward 10-20 mils into the lands. This may help reduce bullet canting in the case neck. The results of the canted bullet test indicate that 0.1 degrees of bullet cant will cause the bullet to be deflected roughly 0.1 inches at 100 yards.

Muzzle Blast Physics

After working on the muzzle blast problem for over a year I still wasn't convinced that I had it pinned down. I had theorized that bullet cant was causing the muzzle blast asymmetry that I had observed in the tuft screen experiment. Also, I had assumed that the muzzle blast asymmetry was pushing the bullet off course. I had shown that bullet cant does occur and demonstrated its effect on dispersion. However, I still wasn't sure that bullet cant was causing muzzle blast asymmetry. So, I decided to try to photograph the muzzle blast flow pattern using spark shadowgraphy. This technique gives you a shadow photograph of the muzzle blast flow. A complete description of the method is in Appendix G. But in brief you use a high energy (10kv), short duration (0.4 microsecond) point spark light source that casts a shadow image either on a piece of film (12"x18" lithograph film) or on a white screen. The image on the white screen can be photographed with Polaroid film. The light rays are absorbed by solid material (bullet, smoke) and distorted by density gradients such as shock waves. Some 100 Polaroid and about 40 lithograph film exposures were made with the bullet at different distances from the muzzle. The flow region between the time the bullet exits the muzzle and outruns the blast wave is called the transitional ballistics region between internal and external ballistics. The shadowgraph studies were done with the rail gun chambered for the 6BR.

Figure 7-13 shows a picture of the precursor spherical shock wave formed by the compressed air and blow-by ahead of the bullet. The bullet is about three inches back down the bore. Figure 7-14 (0 μsec) shows the bullet just emerging from the bore and the precursor shock wave is still ahead of the bullet. One μsec (microsecond) is one millionth of a second. Figure 7-15 (11 μsec) shows the gas escaping behind the bullet when the base of the bullet is about one bullet length out of the muzzle. Figure 7-16 (56 μsec) shows the bullet about 2.5 bullet lengths (12 calibers) from the muzzle and it is about halfway through what is called the Mach disk. If you look carefully you can see a dark vertical line near the base of the bullet. This is a normal shock wave caused by the fact that the gas flow is faster than the bullet (reverse flow). There are also shock waves emanating from the edges of the barrel caused by the expansion of the jet and you can see a conical shock wave forming on the nose of the bullet. Figure 7-17 (90 μsec) shows the bullet in the process of penetrating the main blast wave and just starting to penetrate the precursor shock wave. If you look at the upper part of the main blast wave you can see where

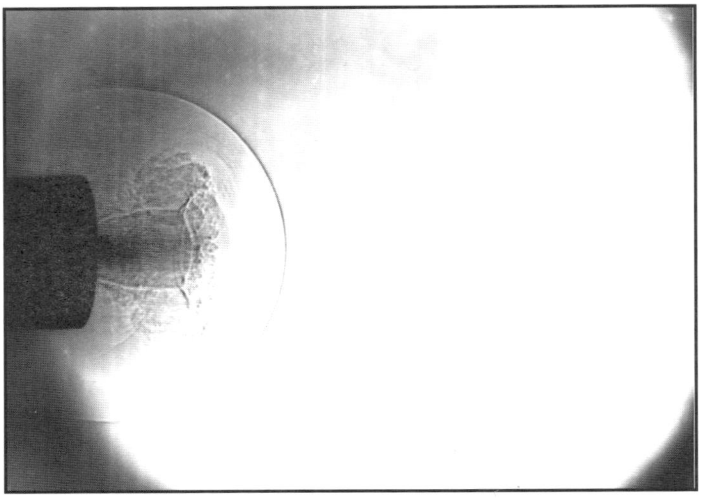

Figure 7-13 - Shadowgraph photo showing the spherical precursor shock wave emerging from the bore. The precursor is formed by the compressed air and gas ahead of the bullet. The bullet is about three inches back in the bore.

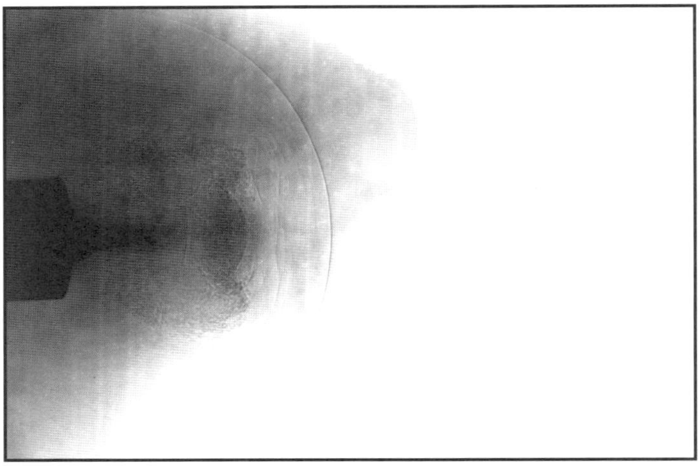

Figure 7-14 - Shadowgraph showing the bullet just emerging from the muzzle at 0 μsec.

Figure 7-15 - Shadowgraph showing the bullet base about one bullet length (0.86 inches) from the muzzle. The spherical blast wave is beginning to form around the opaque cloud of smoke at about 20 μsec. The pressure on the base of the bullet is about 4000 psi.

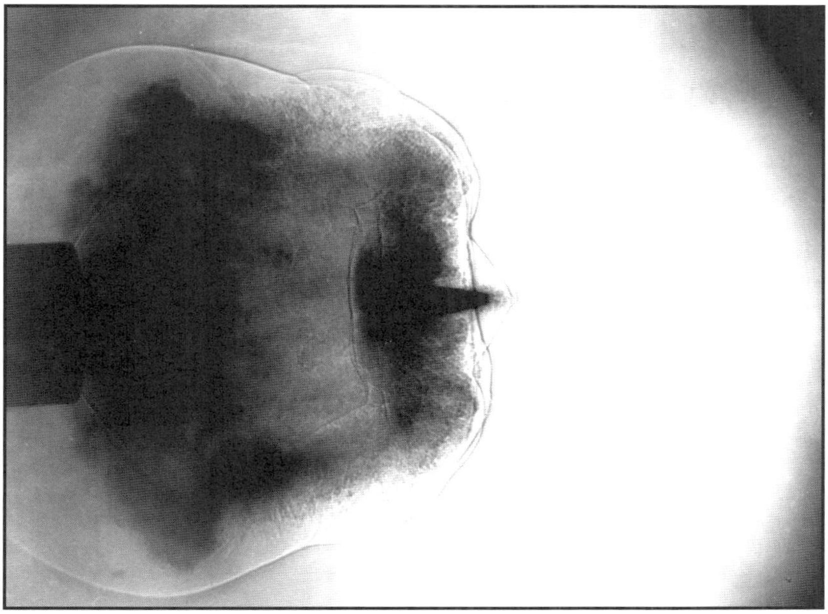

Figure 7-16 - Shadowgraph showing bullet base about 2.5 bullet lengths ahead of the muzzle at 56 μsec. By this time the pressure acting on the base of the bullet has dropped to a few hundred psi. You can see a normal shock wave from the reverse flow on the base of the bullet and bow shock on the tip of the bullet.

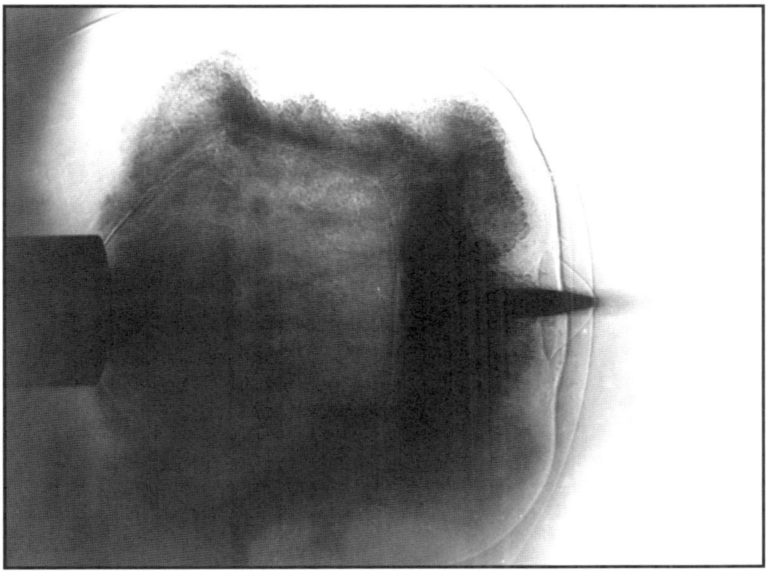

Figure 7-17 - Shadowgraph showing the bullet penetrating the main blast wave and starting to penetrate the precursor at 90 μsec. The main blast wave is overtaking the precursor. The bullet is out of any significant effect from the muzzle jet. There are tiny unburned powder particles trying to penetrate the blast wave above the bullet.

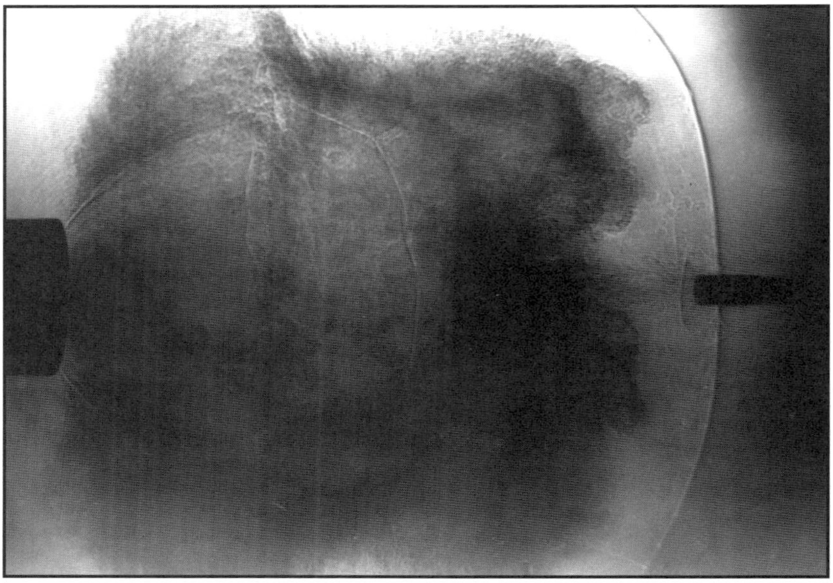

Figure 7-18 - Shadowgraph at 142 μsec showing the bullet penetrating both the blast and precursor waves which have combined. There is a weak bow shock on the nose tip of the bullet.

some unburned powder particles have just slightly penetrated the blast wave. In the last photograph, Figure 7-18 (142 μsec), the bullet has penetrated the blast wave, which has combined with the precursor shock wave and the bullet is leaving the effects of the muzzle blast.

Now, some explanation is necessary. Originally I decided to do this experiment, even though it was time consuming (10 months) and expensive ($1000), to see if I could see the asymmetry in the muzzle blast that I thought was there. Well, after examining over 100 images, most with canted base bullets, **I could see no significant muzzle blast asymmetry like I had expected to see!** While there were occasional minor distortions of the shock waves, there was no orderly, consistent data. This meant that the dispersion observed in the tests with both canted base bullets and canted bullets had to result from some cause other than muzzle blast asymmetry. When you think about it the difference in time between when the short side of the canted bullet leaves the muzzle and when the long side exits is only 0.2 μsec. Not much can happen in that length of time. The other thing that happens is that spherical shock waves behave something like soap bubbles and they try to maintain a symmetrical spherical shape.

So why did the tuft screen experiments indicate that there was muzzle blast asymmetry when we didn't see it in the spark shadowgraphs? Well jets are unstable and perhaps the vortex that forms around the jet isn't always symmetrically located. Also, the vortex ring is similar to a smoke ring and can drift with the wind. At any rate the tuft screen studies, which got me in this mess in the first place were misleading. The only thing I could think of was to try to analyze the observed bullet cant and canted base dispersion data using the 6DOF trajectory simulation computer code (see Chapter 10) to see if that would tell us what causes the dispersion.

Theoretical Analysis

Figure 7-19 shows a sketch of the two ways that testing was done. Figure 7-19(A) shows the canted base test and Figure 7-19(B) shows the situation where the whole bullet is canted (i.e., the real case). In the canted base situation a force vector resulting from the muzzle blast pressure acting on the base is drawn perpendicular to the base. Pressure can only act perpendicular to a surface. This force is under the CG of the bullet producing a moment in the nose up direction. Later we measure the muzzle blast pressure (5000 psi for the 6BR) so we can calculate the force. From some other data that I had

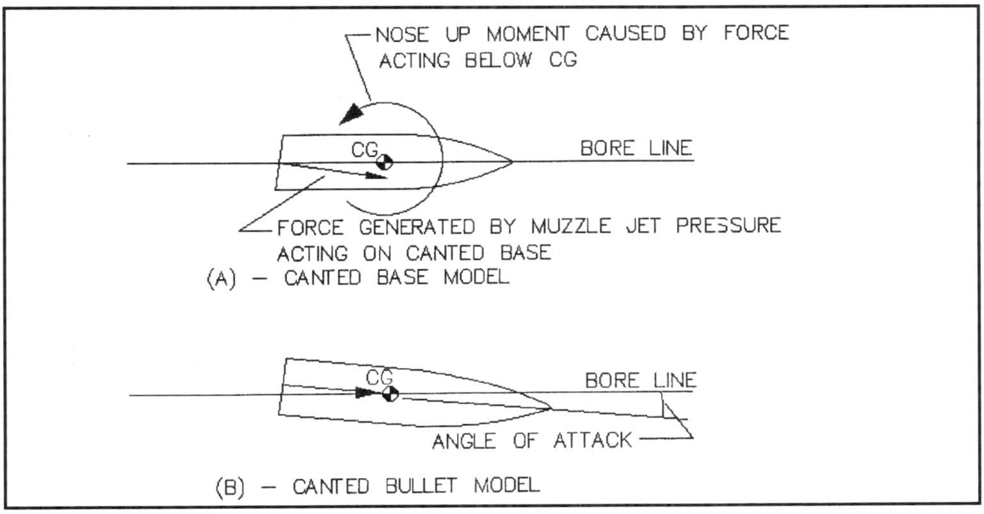

Figure 7-19 - Drawing showing the physical model of the effect of muzzle blast on a bullet with canted base and a canted bullet. When implemented in a 6DOF trajectory simulation computer code the experimental radius of dispersion is correctly predicted.

(Reference 4 and 20) it was clear that once the base of the bullet went through the Mach disk, the pressure on the base of the bullet was very small and could be ignored. From the spark shadowgraphs we can determine the time between when the bullet first emerges and penetrates the Mach disk to be about 50 μsec (8 calibers). Now we know that the muzzle blast pressure drops from 5000 psi to about 150 psi in 50 μsec for the 6BR but we don't know just how the pressure drops off. I had some other data that indicated that the pressure drop was an exponential decay and that assumption was used in the 6DOF computer simulation program. When I ran the computer program it indicated that the radius of dispersion for the 6BR should be about 0.64 inches. If you compare that with Figure 7-11 you can see that the theory agrees with the experimental value of 0.64 inches. I ran the same calculation for the 270 using the measured muzzle blast pressure of 11,500 psi and got 0.85 inches for the radius of dispersion. You can see that the value for the radius of dispersion for the 270 in Figure 7-10 was 0.8 inches, so the theory agrees well with both the 270 and the 6mm experimental data.

Well, I think we now understand the effect of muzzle blast pressure on canted bullets. Some degree of in-bore bullet cant may always occur. Factors that effect bullet cant are a tapered bullet afterbody, the length of the cylindrical afterbody, and how well the bullet is centered with respect to the bore axis before firing. Factors that effect how perfectly the bullet is centered with respect to the bore axis are the amount of case neck run out, the amount of bullet run out in the loaded round, how well the chamber axis is aligned with the bore axis, the degree of throat asymmetry, and the bullet seating depth into the lands. It is also possible that the base of the bullet is not always perpendicular to the centerline of the bullet.

So, how do we minimize the error caused by muzzle blast pressure?

Resized Bullets

I first stumbled onto this trick of resizing commercial bullets purely by accident in the late 60's. I had a Remington 721 chambered for the 300 Weatherby that shot 5-shot groups of 2.5 inches at 300 yards. I then built a similar rifle chambered for a 270 Magnum using the 270 Wby case with a straight shoulder which only shot 3.5 inch groups at 300 yards. I had some theories at the time as to why the 300 shot smaller groups than the 270 that led me to try

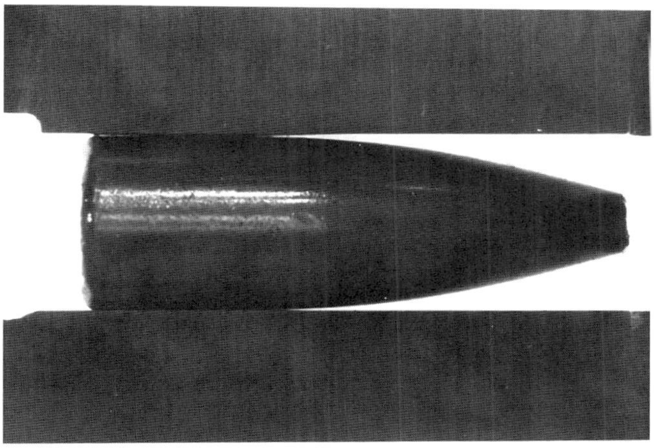

Figure 7-20 - Photograph of a 90 grain 270 HP bullet resized from 0.2770 to 0.2765 inches in diameter held between the parallel jaws of a micrometer. Comparison with Figure 7-6 shows how resizing the bullet increased the cylindrical afterbody length.

resizing the 270 bullets. I tried resizing the 270 bullets by 0.3, 0.5, and 0.7 mils. I found that the bullets resized by 0.5 and 0.7 mils reduced the size of the 300 yard groups from 3.5 inches down to 2.5 inches. Well, it has only recently dawned on me that what was really happening was the effect of muzzle blast pressure on bullet cant. Resizing the 270 bullet had reduced bullet cant. You see, I was using 180 grain Remington Bronze Points in the 300 Mag, which had long cylindrical afterbodies, and I was using 150 grain SP bullets in the 270 Mag, which had tapered afterbodies.

The effect on the length of the cylindrical afterbody of slightly resizing a bullet can be seen in Figure 7-20 which shows a resized 90 grain HP held in a micrometer. Comparison with Figure 7-6 will show how resizing has lengthened the cylindrical afterbody. I also recovered some fired resized 90 grain HP bullets that showed no evidence of canting. Recall that the recovered regular 90 grain 270 HP bullets did show evidence of canting.

I tried resizing 6mm match bullets but it didn't help. All of the 6mm match bullets from three different custom bullet makers that I checked had a thing called a pressure ring at the base of the bullet. Right at the heel of the bullet there is a narrow ring that varies from 0.3 to 0.6 mils over groove diameter, depending on the source. I couldn't understand whether this condition resulted from someone's creative idea or was an artifact of production. So, I called Walt Berger and he said it was an artifact of production and no one seemed to know how it happens. However, the bullets shoot well so they don't try to change it. Incidentally, my handmade 270 flat base bullets don't have this pressure ring so I think it has to do with the relationship between the core swaging die and pointing die diameters. Some folks believe that the pressure ring provides a better gas seal, and maybe it does. However, I noticed that it is comparatively easy to push the point of the bullet off center because the case neck only grips the rear end of the bullet. I don't know whether this is good or bad. This pressure ring may make it easier for the bullet to line up with the throat.

Case Neck Asymmetry

I had hoped to avoid getting into the case neck asymmetry problem, even though bench rest shooters go to great lengths to correct it. The goal is to have the bullet enter the throat perfectly aligned and exactly on center with the bore. Otherwise, the bullet may cant while entering the throat. If you check the run out of the bullet axis with respect to the case axis on factory ammunition you will observe as much as 6 mils eccentricity (total indicated run out, TIR), which is a lot. You can also measure as much as ±1 mil variation in neck thickness around the periphery of the neck, although the variation is usually more like ±0.5 mil. The 270 chamber in the experimental rifle has a neck diameter of 0.307 and a loaded cartridge has a neck diameter of between 0.303 and 0.306, depending on the brand of the case. Case neck average thickness varies between 13.0 and 14.5 mils between brands. This means that the outside of the case neck can be off center by as much as 2 mils. If you add the 1 mil variation in neck thickness, it is possible for the base of the bullet to be at least 3 mils off center. Now even if the bullet is seated well forward into the lands the bullet can be canted by as much as 0.5 degrees. Similar results were obtained on the 6mm Remington. Consequently, case neck asymmetry may have more to do with bullet in-bore canting than the shape of the bullet, and if one is really striving for good accuracy something has to be done to correct this situation.

At first I tried to realign the outside surface of the neck by making a two piece die. It held the case body and shoulder in one piece and the neck in a separate piece. The die was spun in the lathe while a ball bearing tool was pressed against the neck portion. This approach did not work at all, even when the necks of the cases were annealed. Then I dial gaged the necks of a number of fired cases and found that the outside surface of the necks ran true. This was encouraging, since I now had something that was concentric to start with! The next thing I did was to machine the inside of the necks removing just enough to end up with a uniform neck thickness. The shell holder which held the case in the lathe collet was made by cutting a chamber in a piece of bar stock. This was then held on the chambering reamer while the outside of the die was turned to the exact internal dimension of a 3/4 inch lathe collet. Next, an ordinary inside boring tool was used with a light cut at high speed to machine the inside of the case neck while the case was held in the die. With Remington and Norma 270 cases the neck thickness ended up being about 13 mils, which means the average neck thickness was reduced by about 1 mil. The next step was to neck resize the case just enough to hold the bullet, and hope that the inside of the neck would remain on center. Well it worked much better than I had expected. Roughly 40% of the cases had essentially zero bullet run out when loaded, about 40% around 0.5 mil and 20% in the vicinity of 1 mil or less. This is a vast improvement over ordinary ammunition. Unfortunately, I don't know of any way of doing this without the proper equipment. The neck resizing die was made by grinding out the body of the die so that the body of a fired case would just fit into the die. The neck resizing portion was opened up with emery paper from 0.299 to 0.302. In this way the die is perfectly concentric with the lathe spindle. Bench rest shooters use rifles that have undersize chamber necks and the case necks must be turned down before loaded rounds will chamber. The case neck is turned down so that the final neck wall thickness is about 8.5 mils and the variation in thickness is kept to less than 0.1 mil from case to case. The radial clearance between the neck of a loaded round and the chamber neck is usually only 0.4 to 0.7 mils. The advantage to the bench rest approach is that the neck need only be very slightly resized. The disadvantage is that tolerances must be very carefully controlled and only modified ammunition can be used.

Later on I found a better way of reducing the bullet run out, that may be a better long term solution. First, I machined the inside of the neck of a fired case to obtain a uniform wall thickness just like before. Then I used an old rifling head to cut six straight rifling grooves in the neck portion of the neck resizing die. The depth of the rifling cuts were adjusted so that the neck of a resized case would be close to the inside diameter of the chamber neck. The

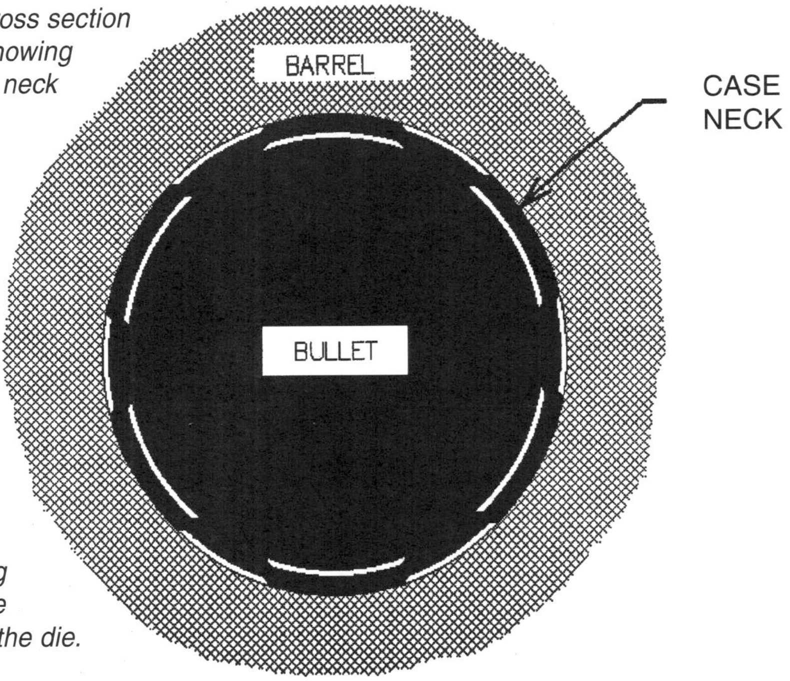

Figure 7-21 - Cross section view of barrel showing how the splined neck cartridge case keeps the bullet centered in the bore. The splined neck case is made by first machining the inside of the neck of a fired case to obtain a uniform neck thickness, then resizing in a die made by making rifling cuts in the neck portion of the die.

case necks are then resized in this die, and the result is a case neck that has a splined appearance. A cross section view of the chamber, case neck and bullet is shown in Figure 7-21. The radial clearance between the case neck and chamber neck is only a few tenths of a mil. You can see that the bullet has to end up on the center of the bore, because there isn't any place else for it to go, if both the outside and inside of the case neck is concentric with the bore. When I checked the run out of the bullets in loaded ammunition, most of the bullets were within a few tenths of a mil with the worst case being about 2 mils, which is a big improvement over doing nothing. Now I realize that everybody doesn't have a precision lathe with collets and a rifling head lying around. However I feel sure that the manufacturers can come up with a cheaper way to do this work. After all, they make special equipment now for bench rest shooters. One advantage of this method is that you don't have to worry about getting a close fit between the chamber and case neck like you do with the traditional bench rest method. You can also work with a standard chamber such as the one in the experimental rifle. However, you would have to seal the tiny gap between the bullet and the inside of the neck to prevent moisture from entering the case, if you are going to store loaded ammunition for an extended period of time. Some people may worry about gas passing through this annular gap. I don't think it's a problem because only a very small amount of gas can travel through such a small radial gap .e. 1.5 to 2 mils). In fact it may prove to be an improvement because it may blow out the burned powder residue between shots.

Just how close this bullet centering with respect to the bore centerline has to be held is not clear. However we know that bench rest rifles won't shoot well without modifying selected cases. This may mean that it is necessary to hold the bullet on center within 0.5 mil or less. Anyway it is time to test fire the experimental rifle with resized bullets and the splined neck cases and see if we can see any improvement. You may recall from Table 4 back in Chapter 4 that the average dispersion was 0.884 inches. The results of this test are shown in Table 8.

TABLE 8

Resized 90 grain 270 HP bullets with splined case necks.

Extreme Spread for eight 5-shot groups at 100 yards

Average	**Maximum**	**Minimum**
0.804	1.399	0.386

Well, as you can see from Table 8 the average dispersion dropped from 0.884 to 0.804. We can use the method of Root Mean Squares (RMS) to evaluate the effect of the resized bullets, which is

$$\text{error} = (0.884^2 - 0.804^2)^{1/2} = 0.367 \text{ inches.}$$

This calculated error agrees well with the estimated error from the previous experimental measurements (0.2 to 0.4 inches). If we make the same RMS calculation on the 300 (2.5 in.) and 270 Mag (3.5 in.) results at 300 yards, that was mentioned earlier, we get an error contribution of 0.81 inch at 100 yards on the magnums. This error is roughly twice that obtained on the 270 Winchester cartridge. This doesn't surprise me, because the muzzle blast pressure on a magnum is 1.5 to 2 times that of a standard case 270. At any rate, I think that the problem of muzzle blast pressure acting on canted bullets can be solved to some extent by using bullets with a cylindrical afterbody of sufficient length and using cases with concentric or spline case necks.

The 270 Winchester did improve some, but not as much as I had expected. Also, with this method of modifying the case necks it appeared that resizing the bullet had little effect, which might be expected. However, the 6mm Rem bench gun accuracy improved dramatically with the splined neck resizing shown in Figure 7-21. It started shooting groups with 6mm match bullets that averaged around 0.25 inches with a 14" twist barrel compared to one inch groups with the old 10" twist barrel. This led me to suspect that something was wrong with the bullets, and it turned out to be due to bullet core failure. We investigate bullet core problems in the next chapter (Chapter 8).

Muzzle Blast Pressure Reduction

The fact that a fast burning powder will result in a lower pressure at the muzzle than a slow burning powder was mentioned earlier. However, for the same muzzle velocity the faster burning powder will produce a higher chamber pressure. Nothing ever comes free in this business! I decided to test this contention by measuring the in-bore pressure at the muzzle using the strain gage method that we used to measure chamber pressure back in Chapter 2. Since we are only interested in the comparison of muzzle pressures resulting from the two different powders, we don't have to go through the tedious calibration procedure used in measuring chamber pressure, and the theoretical calibration will suffice. We will use 49 grains of IMR 4064 and 57 grains of IMR 4831 for the test, which yield approximately the same peak chamber pressure. The 90 grain 270 hollow point bullet was used with both powders. The test results (Figures 7-22 and 7-23) showed that the muzzle pressure was 11,500 psi for IMR 4831 and 7,100 psi for IMR 4064, which means that the muzzle pressure was about 38% less for the faster burning IMR 4064 powder. The muzzle velocity for the IMR 4064 was about 150 fps less than that of the IMR 4831. The negative blip on the oscilloscope data is probably caused by a compression wave that runs a few inches ahead of the barrel expansion caused by the internal pressure behind the bullet. The base of the bullet passes under the strain gage at about 1.37 msec where the peak pressure occurs, and the base of the bullet exits at about 1.40 msec. Notice that the muzzle pressure of 11,500 psi agrees well with the pressure on the base of the bullet at the point of muzzle exit shown in Figure 2-21. So now we know that the choice of powder significantly effects the magnitude of the muzzle blast pressure.

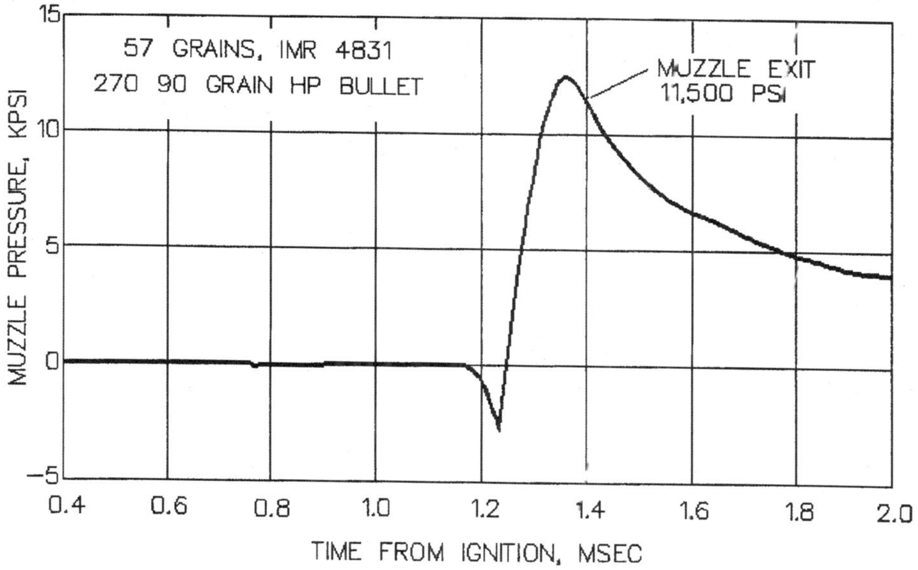

Figure 7-22 - Measurement of the muzzle pressure near the muzzle of the barrel for IMR4831 powder.

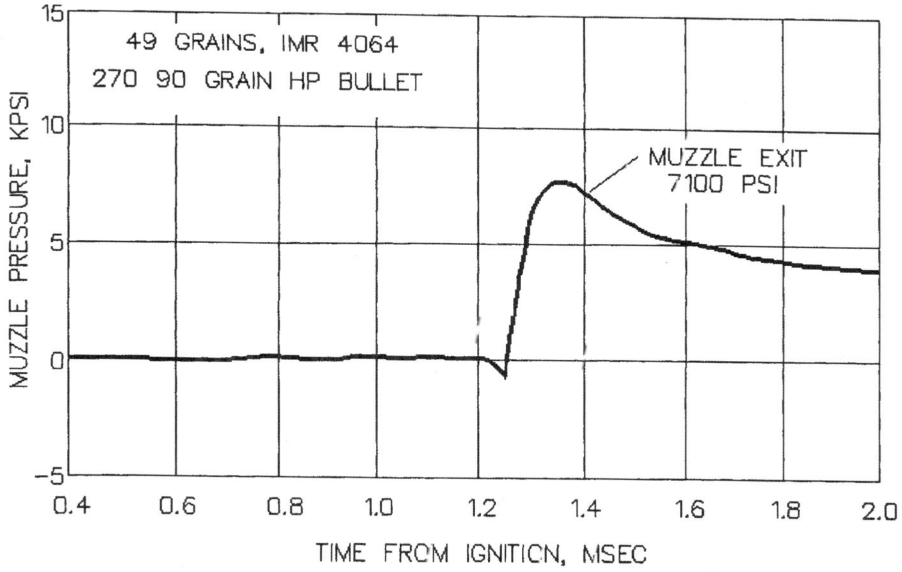

Figure 7-23 - Measurement of the muzzle pressure near the muzzle of the barrel for IMR4064 powder.

Before leaving this fast and slow burning powder discussion, we should point out that the major difference between these two single base powders is in grain size. The grain diameter of IMR 4064 is 0.032 inches and the grain diameter of IMR 4831 is 0.041 inches. This means that IMR 4064 will burn out about 28% faster than IMR 4831 if the burning rate is the same for the two powders. This means that the peak chamber pressure will occur at about 0.47 msec for IMR 4064 compared to 0.65 msec for IMR 4831 (see Figure 2-21, Chapter 2). I am told that all IMR powders are single base powders, which means that their burning rates are similar, and the main difference is in grain size. The burning rates of all single base gun powders are roughly the same, regardless of whether you are working with rifle powder or 8 inch cannon powder. The burning rate can also be modified by the addition of inhibitors. We can observe the effect of grain size in Table 9, where the grain diameter and web thickness is shown for several powders that range from slow to fast burning.

TABLE 9

Powder Grain Diameter

	Powder	Grain Diameter (inch)	Web Thickness
"Slow"	H 570	0.057	0.0285
	H 4831	0.045	0.0225
	IMR 4831	0.041	0.0175
	IMR 4350	0.038	0.0160
	IMR 4320	0.034	0.0140
	IMR 4064	0.032	0.0130
	IMR 3031	0.029	0.0115
	H 322	0.027	
"Fast"	IMR 4198	0.026	0.0100

This table demonstrates the large difference in grain diameter and web thickness between fast and slow burning powders. The data were obtained by actually measuring the grain diameter with a micrometer, and are slightly different from Du Pont published data in a few instances. Consequently, when we talk about fast and slow burning powders, we are generally talking about fine and coarse grained powders, and the time it takes for the grain to burn.

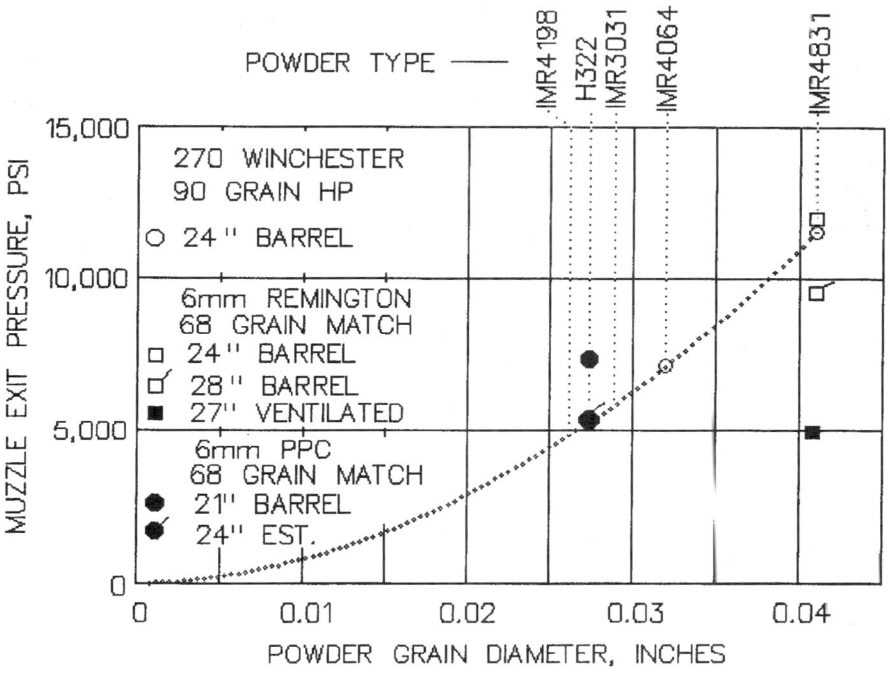

Figure 7-24 - Graph showing how powder grain diameter effects muzzle pressure. The larger grain slow burning powders produce a higher pressure at the muzzle than smaller grain faster burning powders.

The two experimental muzzle pressure points for the 270 (open circle symbols) are plotted in Figure 7-24, and a curve faired through them. This allows one to graphically see the effect of grain size on muzzle pressure. For instance, changing from IMR 4831 to IMR 3031 powder will reduce the muzzle pressure by nearly a factor of two.

So now let's talk about the other data points on Figure 7-24. The square data points are from the 6mm Remington in the rail gun. You can see that the muzzle pressure for the 270 Winchester and 6mm Remington both with 24" barrels is practically identical for IMR4831. It should be, because the ratio of case capacity to bore area is about the same. The next thing to notice is that the muzzle pressure for the 6mm Remington with a 28" barrel is about 9,000 psi. This pressure is about 4,000 psi less than it is on the 6mm Remington with a 24" barrel.

Now notice the black square at about 5,000 psi which is the muzzle pressure for the 6mm Remington with a 27" barrel that has a ventilated muzzle (see

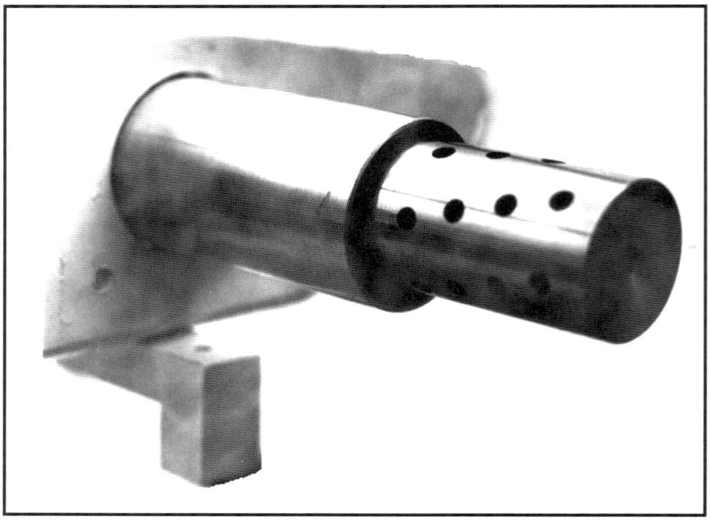
Figure 7-25 - Photograph of the ventilated muzzle.

Figure 7-25). The muzzle is ventilated with twenty four 0.078" diameter holes in the bottom of the rifling groove. The rifling lands are uninterrupted and continue to support the bullet until muzzle exit. The larger outside holes are 3/16"D drilled within 0.15" of the rifling groove. This allows insertion of a 5/64"D milling cutter with a 3/16" shank to cut the final 0.15 inches. A milling cutter was used to cut the holes to minimize the formation of burrs. The inside sharp edges of the vent holes burn off after a number of rounds are fired (perhaps 50) and they do not disturb the bullets. By the way the groove is 0.093" wide so the hole is slightly smaller than the groove. The bullet is constrained by the lands until it exits the muzzle and the pressure at the muzzle is much reduced. Do not confuse this method of muzzle venting with the usual recoil reducer. The usual recoil reducer uses a counter bore with a diameter greater than the bullet diameter. In this case the holes do not have to be precisely located in the groove. However, the bullet is upset the instant it leaves the constraint of the lands and even though the muzzle pressure is reduced it can't reduce the effect of muzzle blast pressure on the bullet exit dispersion.

I was also able to measure muzzle pressure on a custom 6mm PPC Light Varmint (LV) bench rest gun owned by a friend of mine (Dr. Jack Jackson). This gun had a 21" barrel which seems to be typical of bench rest rifles and the data is shown by the black circle symbol. The load was a fairly hot load of H322 (27.8 grains), which is a fast burning powder. Consequently, the peak chamber pressure is probably in the vicinity of 60,000 psi instead of the 53,000 psi chamber pressure for the 270 and 6mm Remington. Therefore the muzzle blast pressure on the 6PPC would be expected to be correspondingly higher. When one corrects the data for a 24" barrel length (flagged black circle symbol), you can see that the muzzle blast pressure is close to the normalized curve. So you see that powder burning rate (grain diameter), peak chamber pressure, and barrel length all effect the muzzle blast pressure. Even so, the 6PPC has a low muzzle pressure by virtue of its small case capacity, which allows the use of fast burning powders with a full case load.

Well OK, did the muzzle ventilation help? It did. It reduced the average 5-shot 100 yard group size from about 0.35" to 0.23" with match bullets and a 14" twist in the 6mm Remington rail gun. This is not as good as a top flight 6PPC HV gun or the 6BR rail gun which average in the high ones (i.e., 0.18"). The ventilated muzzle might be a good idea on sporters and long range magnum rifles, but there is talk about outlawing it in Hunter Class bench rest competition because of the increased muzzle blast on nearby shooters. I personally can't tell if it makes much difference.

The vented muzzle was sectioned and the sharp corners at the vent holes had been rounded on the downstream corner of the holes by the hot gasses. Also there were copper smears just downstream of the vent holes. Just how important this is I don't know, but it can't be good. This type of muzzle venting is not a trivial machining job and would be expensive to do. Indexing the vent holes so that they end up in the grooves is difficult.

What this all amounts to is that the effect of muzzle blast pressure on group size can be decreased by reducing the muzzle blast pressure. The muzzle blast pressure can be reduced by using a smaller grain or faster burning powder, longer barrels, lower chamber pressure, or a muzzle ventilator. However, for a given case volume you have to give up velocity, or bullet weight, or increase the maximum chamber pressure to obtain the same velocity. This usually means that one has to use a lighter load with the faster burning powder that won't fill the case, which may result in greater shot-to-shot velocity variations.

The bench rest shooters optimized this problem by going to the 6 mm PPC which has a much smaller case than the regular 6 mm Remington, and using powders that are as fast or faster than IMR 4198 with 60-70 grain bullets. Consequently, the muzzle blast pressure effect is greatly reduced. Unfortunately, while this works well at moderate ranges, the light bullets with low sectional densities don't do well at long (600-1000 yard) ranges. So, at long ranges the only thing that can be done is to use heavy bullets and larger capacity cases with slow burning powder. These heavy long range target guns usually have long (30 inch) barrels that reduce the muzzle blast pressure to some extent.

The 270 Winchester cartridge that we are working with is a medium capacity case that is filled with IMR 4831 or IMR 4350. These powders provide optimum velocities, but also give high muzzle blast pressures. The solution is to use slow burning powders with heavy bullets (130 grain) for hunting and to switch to a faster burning powder and light bullets (90-100 grain) for target shooting.

Muzzle Crowning

Just how one shapes or crowns the muzzle of a barrel for best results has been the subject of an endless series of articles in print over the years. As far as I can see most of this stuff has been pure drivel (unadulterated by any facts). I think the only fact that we have to go on is that nobody really knows, including this author. As far as I can tell it doesn't really matter, as long as the crown is symmetrical and perpendicular to the bore. I've tried all of the most common shapes (Figure 7-26), and I can't tell any difference between them. The circular or sporter crown, which is used on most commercial guns, looks nice and does a good job of protecting the end of the rifling. However, it is more difficult to machine and keep centered than the two other types of crowns shown in Figure 7-26. The bench rest flat crown is the easiest to machine and least sensitive to just how well the bore is centered. It may be slightly indented near the bore to protect the rifling, although I just use a flat crown without the indentation on target rifles. Many custom bench rest barrels have the 11 degree conical crown. Why this should be any better, I don't know. However, the 11 degree angle probably came from the fact that tapered afterbodies (such as boat tails on bullets) will suffer from flow separation if the cone angle is greater than 11 degrees. This has nothing to do with jet flow. Someone probably saw this information and incorrectly assumed that it would apply to muzzle jet flow. But the conical crown is probably as good as any, and maybe they know something I don't. The bottom line is that nobody really knows what crown shape is best, or even why it might be best. Consequently, you might as well use whatever crown that you like, as long as it is symmetrical with the bore axis.

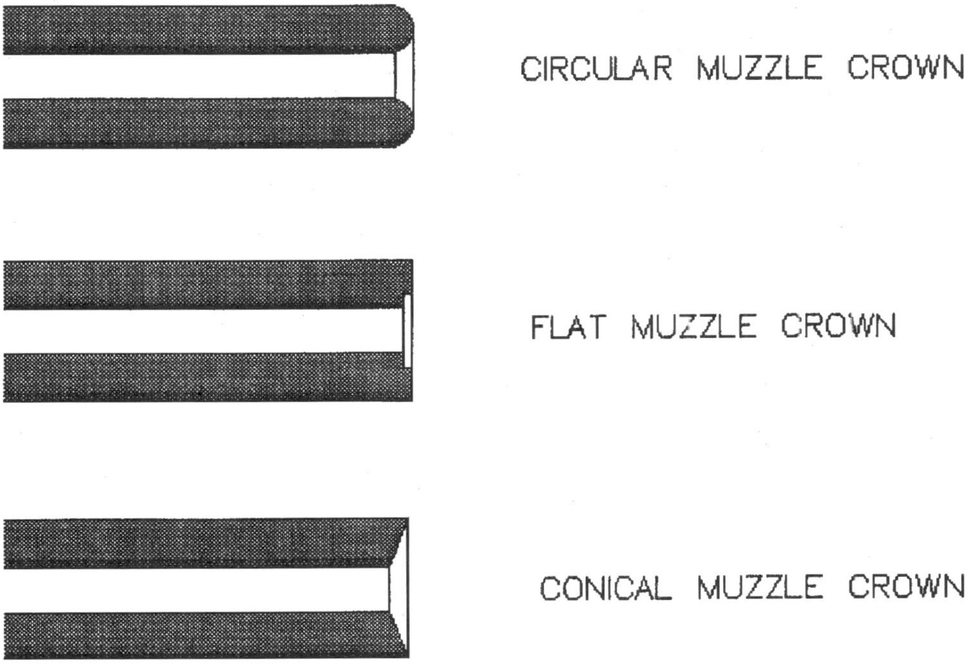

Figure 7-26 - Various techniques of crowning gun muzzles.

Summary

By using the tuft screen we found that the muzzle blast appeared to be asymmetric. We found by inspection that most bullets do not have a cylindrical afterbody of sufficient length to prevent the bullet from canting in the bore. Bullets were recovered which showed canting, even though the measurement accuracy was less than desired. The measurements taken on the recovered bullets indicated a cant angle of 0.25-0.5 degrees.

Bullets were then modified by slicing off the base at a two degree angle and these bullets were then fired in four groups with the canted bases indexed every 90 degrees. The result showed that a two degree base cant produced a radius of dispersion of 0.8 inches for the 270 and 0.64 for the 6mm BR at 100 yards. Later a computer code was developed which accurately predicted the test results.

We then took spark shadowgraph pictures of the flow field in an effort to detect asymmetries in the muzzle gas flow. Much to my surprise, no significant flow asymmetry was observed. This observation led to the development of a computer code that accurately predicted the canted base target results for both the 270 and 6mm bullets. Firing tests were then conducted with the bullet canted in the case neck at an angle of 0.215 degrees. These tests showed a radius of dispersion of 0.196 inches at 100 yards and the computer code predicted a radius of dispersion of 0.243 inches. Since 0.2 degrees of bullet cant can easily happen, the dispersion from canted bullets can be large.

We then explored methods of reducing bullet cant. The 270 bullets resized in diameter by 0.5 mils, that were recovered, demonstrated much less in-bore bullet cant. Resized bullets reduced the average group size from 0.884 to 0.804 inches at 100 yards in the 270, indicating that bullet in-bore cant was corrected to some extent by resizing. Resizing 6mm match bullets had no effect on group size in the 6BR. The fact that resizing the 6mm match bullets had no effect was likely caused by the case neck machining combined with seating the bullets into the lands which helps prevent significant bullet canting. The bullets were found to be off center in unmodified 270 and 6mm Remington cases and this was corrected by machining the inside of the case necks, and using a special spline crimp. The spline crimp greatly improved the accuracy of the 6mm Remington with Cook match bullets, but only had a small effect on the 270. At this point it became obvious that something was wrong with the 90 grain 270 bullets that were being used, and we investigate this in the next chapter.

The muzzle blast pressure was measured using strain gages and it was determined that large cases (270 Win.) with relatively slow burning, large grained powder had a much larger muzzle blast pressure than relatively fast burning small grained powder. Muzzle blast pressure was also decreased with longer barrels and ventilated muzzles. Muzzle ventilation was tried and it did reduce the muzzle blast pressure as expected and produced a significant reduction in group size. However, the type of muzzle venting that was used is a difficult machining job that would be expensive to do in production.

CHAPTER 8
BULLET CORE PROBLEMS

The lead core in a jacketed bullet is subjected to a large shearing stress at the interface between the jacket and the core during spin-up. As the bullet enters the rifling a large angular acceleration occurs which spins up the jacket. The lead core is heavy and has a large spin moment of inertia that resists this large angular acceleration. The core is driven by friction forces between the core and the jacket and shear stresses developed by the internal indentations in the jacket caused by the rifling engraving. These internal indentations protrude into the core about two mils. If the lead core is too weak to stand this shearing stress, core stripping results and the core will have a slower spin rate than the jacket when the bullet exits the muzzle. The maximum differential spin rate that I measured (5.5%) results in the core roll angle lagging behind the jacket roll angle by as much as 20 degrees. After muzzle exit the core slows down the jacket spin rate and the jacket speeds up the core spin rate slightly until both the core and jacket are at the same spin rate. The resulting spin rate of the bullet is slower than it would have been if core stripping had not occurred. Just how this effects the bullet's trajectory is not known. However, it probably results in a center of gravity (CG) asymmetry and certainly produces a slower, variable spin rate. We can, and will measure the variation in spin rate, but I don't know of any way to measure the effect of core stripping on CG asymmetry. We start out by determining core hardness as a measure of the shearing strength of the lead cores in several bullets. We also measure the torque required to strip the core in various bullets. This will tell us how likely it is for core stripping to occur.

Figure 8-1 - Photograph of the Brinell Hardness Tester.

Laboratory Core Stripping Tests

The first thing that was done was to make a Brinell Hardness Number (BHN) test device which is shown in Figure 8-1, because I couldn't find a local laboratory that had one. The BHN test device is nothing more than a spring loaded plunger that screws into a loading press and applies a known load on a small (3/16" diameter) ball bearing, which creates a small crater in the lead sample. If the load, ball diameter, and crater diameter are known, the BHN can be determined from the following equation.

$$BHN = 0.0004485*F/\{(\pi/2)*D^2*[1-\sqrt{(1-(d/D)^2}]\}$$

where

F = load, pounds (100 typical)
D = ball diameter, inches (0.1875)
d = diameter of crater in sample, inches
$\pi = 3.14159$

According to engineering handbooks and the experimental stress tests that I ran in my hydraulic press on several samples, the core yield stress, or strength of the core can be determined by multiplying the BHN by 515. The results of these hardness tests for bullets from four different manufacturers are shown in Figure 8-2. The data for pure lead and Linotype metal are included for comparison. You can see that the measured core hardness and strength varies a lot between bullets—much more than I would have expected. I chose these bullets to work with because I had reason to believe from test firings that the 65 grain 6mm match bullet never strips at a moderate load in a 10 inch twist, that the 270 90 grain HP bullet is marginal in a 10 inch twist, and that the 68 grain match bullet always strips in a 10 inch twist. Now before anyone gets excited I want to point out that the 6mm 68 grain match bullets perform very well in bench rest rifles with a 14 inch twist, where they are intended to be used. So, we can see that the core hardness and strength results are in qualitative agreement with the firing test results. So, what does this prove?

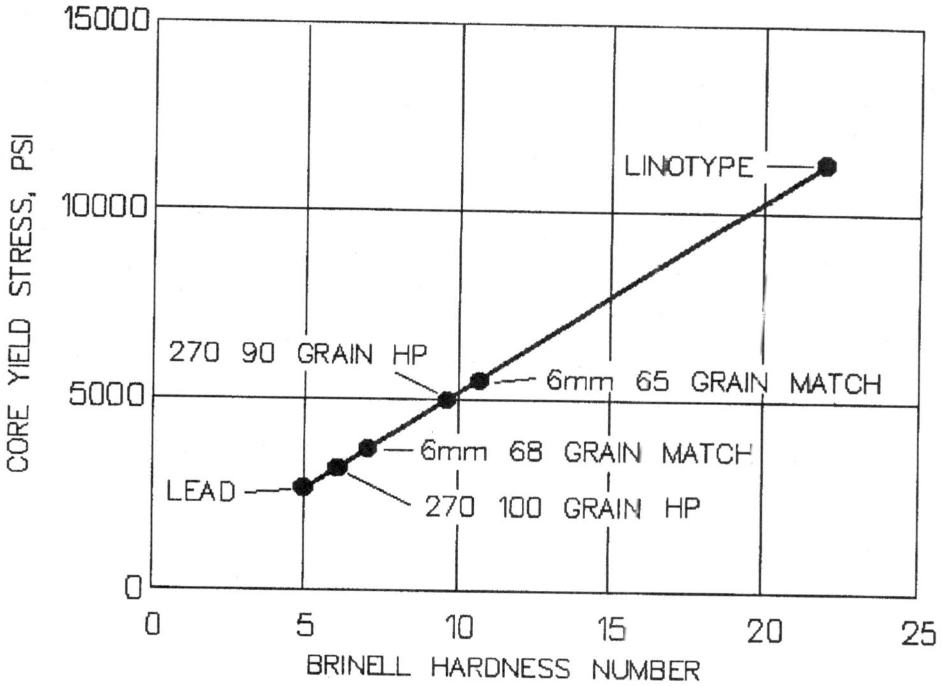

Figure 8-2 - Measured core Brinell hardness and core yield stress (strength) for the cores of four test bullets. Pure lead and linotype metal are added for reference.

It proves that the 65 grain match bullets have the strongest core material in the lots of bullets tested. However, it doesn't tell us how much torque it takes to strip a core. To find out, we will have to measure it.

Figure 8-3 shows a torque wrench designed to measure the core failure torque. A blade similar to a screw driver extends down into the lead core from the end of the torque wrench. The blade of the wrench is inserted into the nose of a bullet that has been swaged into a short section of barrel. The top 1/8th inch diameter rod serves as a pointer and the bottom rod serves as a flexure, and is used to apply a torque to the driver blade. The torque required to cause core failure is read on a calibrated card attached to the flexure rod. The assembly is held in a vise with the pointed end up. A downward compression load of about 1000 pounds is applied to the pointed end, which partially simulates the set back load. The top of the device is pointed to reduce the friction torque between the top of the wrench and the press that is applying the simulated setback load. A 3/16 inch diameter flexure rod was used in testing 270 bullets. The results of the core failure torque are shown in Figure 8-4 for room temperature (70 °F) and at an elevated temperature of 250 °F.

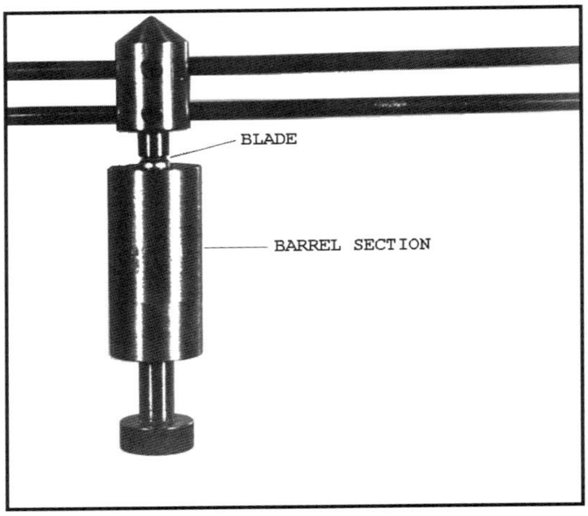

Figure 8-3 - Photograph of the torque wrench used to measure the core failure torque on four test bullets.

Now I don't know for sure just how hot a sporter bore gets. I know that the bore gets above 200 °F on a hot day after firing several shots, because water boils when I pour it into the barrel to cool it off. Also, bore surface temperatures of 1000 °F of short duration have been measured by the Army (Reference 3). As you can see, temperature makes a big difference, and I suspect that this is one of the reasons why hot barrels usually don't shoot very well. I have measured chamber temperatures of 133 °F on a cool day after firing 15 rounds, and I would expect the throat temperatures to be significantly higher. I don't know how hot the bullet gets sitting there in the throat for a while, but it must be over 133 °F on a hot day after one fires several rounds. Also shown on Figure 8-4 (dashed lines) is the estimated

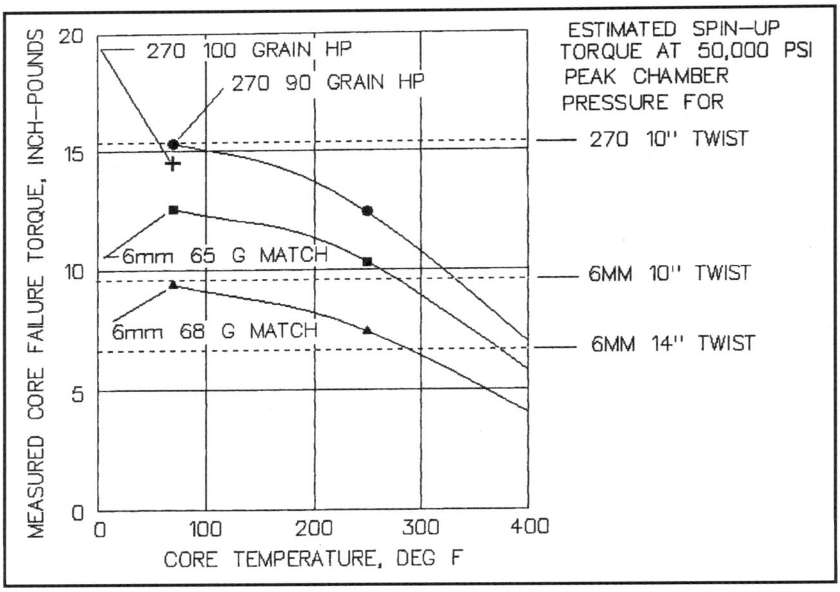

Figure 8-4 - Core failure torque for the four test bullets at normal and elevated temperatures.

spin-up driving torque for the 270 with a 10 inch twist and for the 6mm with either a 10 or a 14 inch twist. You can see that the 65 grain match bullet core should survive spin-up while the 68 grain match and the 90 grain 270 bullets should be either marginal or fail in a 10 inch twist. The 68 grain 6mm bullets should survive a 14 inch twist and the experimental data shows that it does. The estimated spin-up torque was obtained by first computing the equilibrium driving torque from the peak chamber pressure, which can be done very accurately, and then multiplying the equilibrium driving torque value by a factor of two to account for the fact that the spin-up torque is a dynamic rather than a static load. The equilibrium angular acceleration is

$$a = (P*A*g*2*\pi)/(W*Tw), \text{ rad/sec}^2$$

where

P = peak chamber pressure, psi
A = bore cross-section area, in^2
W = bullet weight, pounds, (grains/7000)
Tw = twist, inches/12
g = gravitational acceleration, 32.2 ft/sec^2
π= 3.14159

The equilibrium driving torque is obtained by multiplying the angular acceleration by the core spin moment of inertia.

$$Te = a * Ix * 12, \text{ inch-pounds}$$

where the spin moment of inertia is

$$Ix = 1/2 * m * r^2, \text{ slug-ft}^2$$

and

m = core mass = (core weight in pounds) / g
r = core radius, ft

The dynamic driving torque (effective spin-up torque) is then obtained by multiplying the equilibrium driving torque by a factor of two. Now, while I know from experience that the dynamic factor of two is quite reasonable for this case, I am unable to quote restricted references, so you will just have to take my word for it. Also note that the spin-up torque is directly proportional to peak chamber pressure. Also, the spin-up torque is, to some extent, a function of the amount of bullet free run before striking the rifling. In this case the free run is short enough that the factor of two is valid. Consequently, our core failure torque measurements and estimated spin up torque boundaries should be reasonably good, and it is likely that the cores are failing in some cases. It should be remembered that the calculated spin up torque boundaries (dashed lines in Figure 8-4) are only approximate and may be off as much as 20%.

The results of this work showed that the core failure torque was directly proportional to the core yield strength on bullets of similar shape. I think it is obvious that the length of the rifling engraving on the side of the bullet will also effect the core failure torque. The core failure torque was also significantly effected by the core temperature. The fact that the core failure torque is sensitive to the length of the rifling engraving may explain why some boat tail bullets do not perform well in some rifles. In general, a flat base bullet will have longer rifling engraving than an equivalent boat tail bullet.

I finally decided that the only way to prove that core stripping was real, was to measure the spin rate of the bullet after it leaves the muzzle, and compare the measured value to the spin rate calculated from twist rate and measured muzzle velocity.

Bullet Spin Rate Tests

If the bullet spin rate is significantly less than the rate determined by the muzzle velocity and the barrel twist rate, then either core or jacket stripping has occurred. Since there was no evidence of jacket stripping on the recovered 270 bullets, any reduction in spin rate must be caused by core stripping. I first tried using an optical detection device, but after several months of unreliable results I decided to use a magnetometer.

The magnetometer device used to measure spin rate is shown in Figure 8-5, and it is nothing more than a square tube measuring 1" square (inside dimension) by 3 feet long. The tube is constructed of 1/4" thick plywood. It has a 1 inch hole in each end for the bullet to pass through. A rectangular coil 1.5 X 1.5 X 31.25 inches is wound on the coil form as shown in Figure 8-6. Holes 1/8" in diameter are drilled through the form for two 1/8" wooden dowels to facilitate winding the coil lengthwise on the form. The coil consists of 35 turns of #30 ASWG magnet wire. The wire is wound on the form by starting at the end of one of the dowels and stringing the wire down one side of the

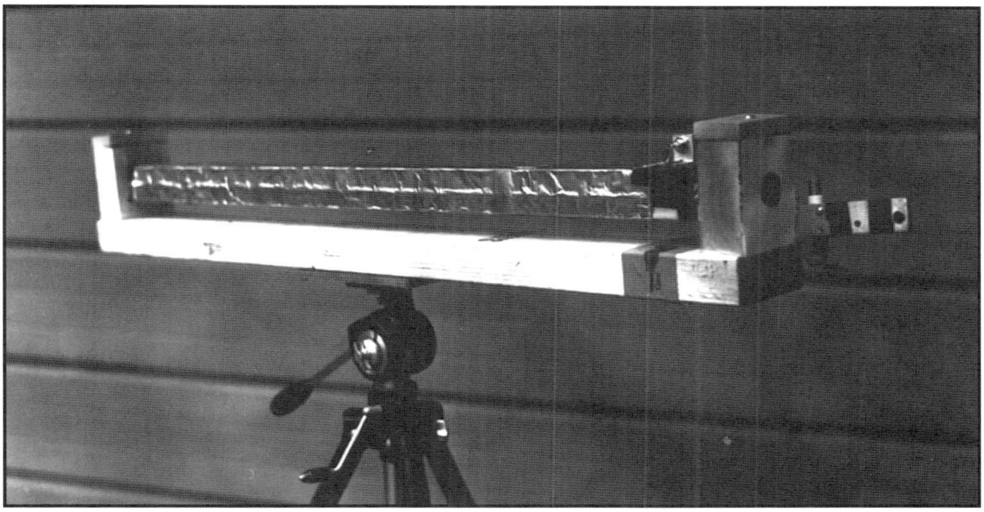

Figure 8-5 - Photograph of the magnetometer device used to measure bullet spin rate after the bullet leaves the muzzle.

form to the dowel at the other end. Then the wire goes over the top of the form to the other end of the dowel and back down the other side. This procedure is repeated 35 times to obtain the coil. A 15 ohm resister is connected in series with the coil and a 1.5 mfd condenser is connected across the output. The condenser reduces the RF noise and the resister provides 0.7 critical damping. A shield of aluminum foil is wrapped around the coil to further reduce electromagnetic noise. The signal to noise ratio in an extreme RF environment is about 200, so a very clean signal is obtained.

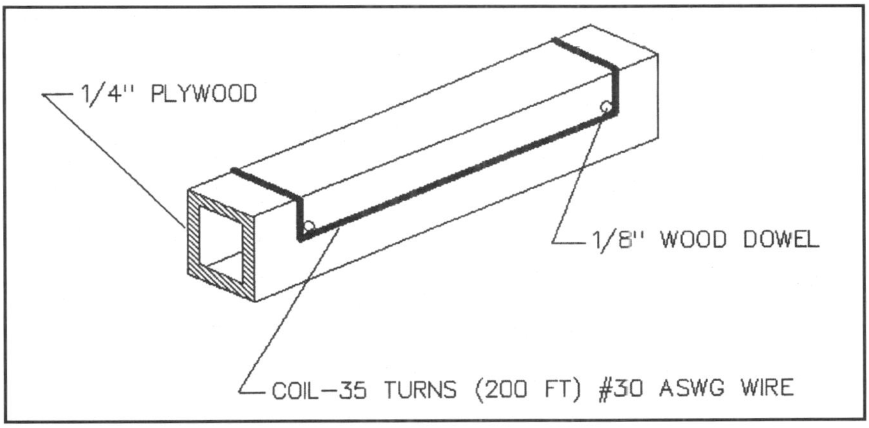

Figure 8-6 - Drawing of the magnetometer showing how the coil is wound on the 1/4 inch plywood form. The coil form is three feet long and the square hole for bullet passage is one inch square.

Figure 8-7 - Photograph of a 270 bullet with the magnet inserted in the 1/16 slot milled into the nose. A rare earth magnet is shown below the bullet.

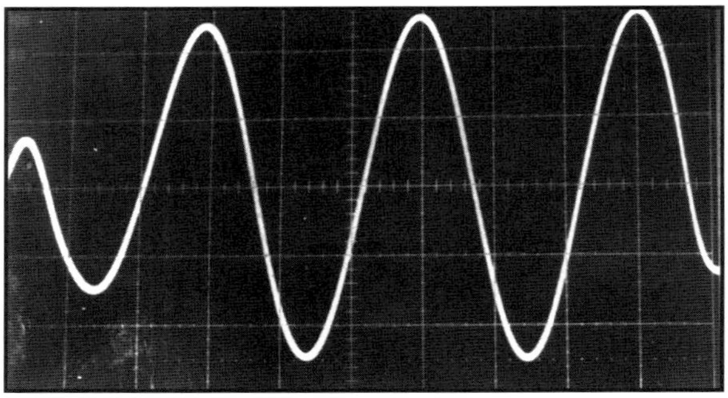

Figure 8-8 - Photograph of an oscilloscope trace showing the sine wave signal generated in the magnetometer coil by the rotating magnet in the nose of the bullet. Horizontal time scale is 0.1 msec per cm. Voltage scale is 0.1 volt per cm.

The magnetometer is placed with its center about 12 feet from the muzzle. An Oehler 35P chronograph with 6 foot screen spacing is placed between the gun muzzle and the entrance to the magnetometer. A small microphone is placed just ahead of the entrance to the magnetometer to trigger the oscilloscope sweep circuit. The bullet has a small (3/16" diameter by 1/16" thick, Radio Shack 65-1895) rare earth magnet epoxy bonded into a 1/16" slot cut in the bullet nose. When the bullet passes through the coil, the rotating magnetic field produces a sine wave electrical signal (0.5 v peak to peak). A photograph of a 270 bullet with the magnet inserted in the nose is shown in Figure 8-7 along with a separate magnet. The electrical signal is displayed on an oscilloscope and photographed. A typical record is shown in Figure 8-8, where you can see that there are about 2.5 complete cycles of data. The distortion of the first half cycle is due to the transient response of the circuit and is ignored. The period of the oscillation can be measured to about 0.2% with this method. The gun is mounted on a machine rest and bore sighted, so that the bullet will pass through the one inch diameter entrance and exit holes in the magnetometer.

The method of testing is to fire a solid copper Barnes X bullet with a magnet in its nose with every five test bullets. Because the Barnes X bullet is solid copper it has no core to strip. The solid copper bullet serves as a reference for the data obtained on the test bullets. The reference bullet tells you what the spin rate should be for a given muzzle velocity and is compared directly to the spin rate measured on the test bullet. A small correction is made for the effect of the small differences in velocity between the reference and test bullets. Since the accuracy of the velocity measurements is no worse than 0.2%, the total error involved in the measurement can't be more than 0.4% in comparing the solid copper bullet with a jacketed lead core test bullet. When the core strips the core will have a slower spin rate than the jacket when the bullet exits the muzzle. After muzzle exit the core slows down the jacket spin rate and the jacket speeds up the core spin rate slightly until both the core and the jacket are spinning at the same rate. The bullet then passes through the magnetometer and the slower spin rate of the bullet that has stripped its core is measured. The measured spin rate difference can be multiplied by about 1.5 to estimate the true spin rate difference between the jacket and core when the bullet exits the muzzle. This is due to the core being much heavier than the jacket. Therefore, its spin moment of inertia is greater than the jackets spin moment of inertia. Therefore the slower spinning core will slow the jacket more than the faster spinning jacket accelerates the core after the bullet exits the muzzle. I have measured differences in spin rate between the reference and test bullets with the magnetometer ranging from 0.0% to 5.5%. There can be no doubt that some of these test bullets were stripping their cores. The difference in roll angle between the core and the jacket can be as much as 20 degrees. In the worst case the spin rate of the jacket, when the bullet exits the muzzle, would be about 8.25% higher than the spin rate of the core (5.5% * 1.5 = 8.25%). This is a very significant difference.

The magnetometer spin rate results verified the core failure torque measurements shown in Figure 8-4 – that is the bullets that were predicted to core strip did strip. A summary of the results of some 61 measurements is shown below for 10 inch twist barrels. There were five or more records for every bullet and the bullet core was considered to have stripped if the difference between the predicted and measured spin rate exceeded 1%.

TABLE 10

Core Stripping Bench Tests

Bullet Type	Load IMR4831	Pressure kpsi	Percent Failure
270 Winchester - 10" twist			
270 90 gr HP	57	53	60%
270 90 gr HP(hot)	57	53+	100%
270 90 gr HP	59	63	100%
270 100 gr HP	57	53	40%
270 100 gr HP	59	63	100%
6mm Remington - 10" twist			
6mm 68 gr match	44	50*	100%
6mm 68 gr match	46	60*	100%
6mm 65 gr match	44	50*	0%
6mm 65 gr match	46	60*	30%

+ indicates pressure higher than shown
*estimated

The elevated temperature tests were run by soaking the bullet in a loaded round in a pan of boiling water, then firing as quickly as possible. The temperature of the bullet probably was around 180°F when fired. The 6mm pressures were estimated from the 270 data in Chapter 2. This spin rate data, combined with the core stripping data tells me that some light hollow point bullets have a tendency to strip their cores in a ten inch twist barrel. It is also likely that some boat tail bullets suffer from core stripping. There has been a lot of discussion in Precision Shooting magazine (1993-1994) about how some rifles are inaccurate with boat tail bullets. Unfortunately, the spin test shot groups in a 10 inch twist barrel average over 1 inch at the same muzzle velocity. I also know that if you push the 65 grain match bullet too hard in a 10 inch twist the accuracy deteriorates. Table 10 shows that the 65 grain match bullet will core strip if it is pushed too hard. I believe that these results tell us that core stripping is significant.

Many bench rest shooters load their 6mm PPC rifles with very heavy charges which results in high chamber pressures (>65,000 psi) and high muzzle velocities (>3300 fps). If you consult Figure 8-4, you will note that the spin-up torques were calculated for a chamber pressure of 50,000 psi. The calculated spin-up torque for the 6mm in a 14 inch twist will move upward by 30% for a chamber pressure of 65,000 psi. This moves the dashed line for the 14 inch twist to where it is just below the dashed line for the 10 inch twist. In other words, if you drive these match bullets too hard you may experience core stripping even in a 14" twist barrel resulting in the occasional flyer that defies explanation.

One other interesting bit of information was obtained as a by-product of the bullet spin rate tests. If we compare the measured spin rate with the spin rate calculated from the barrel twist rate and the measured velocity, we find that the measured spin rate is less than the spin rate obtained from the measured velocity by about 2.7%. In one sample case where the measured velocity was 3057 fps, the muzzle velocity as determined from the spin rate was 82 fps less than the measured velocity after correction for chronograph distance from the muzzle. This phenomena was described in Chapter 2, and is due to the muzzle jet continuing to act on the base of the bullet after the bullet leaves the barrel. The muzzle jet accelerates the bullet after it leaves the muzzle but the spin rate is not increased because the bullet is no longer in contact with the rifling.

Core Collapse

Unfortunately, the 270 bullet that we chose for this investigation probably suffers from core collapse, however I doubt that this is a common flaw in bullet design. Figure 8- 9 shows a photograph of the 270 90gr HP bullet of the same type that we have been using. You can see that the core projection is about 1/8 inches in diameter and extends about 0.15 inches forward in the nose of the bullet. Now a quick calculation will show that with a peak chamber pressure of 50,000 psi the compressive stress acting at the base of the core projection, due to the setback acceleration, is approximately 12,000 psi compared to the yield stress of the lead of 5,000 psi that we measured (see Figure 8-2). Obviously, the core projection will fail because the applied stress is nearly three times the yield stress of the core. If the core projection stays axially symmetrical during collapse, it probably won't effect accuracy very much. However, if it slumps off to the side, it will cause a principal axis and CG asymmetry, which do effect accuracy as we will see in Chapter 9. One could prove conclusively that the core collapses by using a softer recovery of fired bullets than was used in Chapter 7, but that would be a lot of work and I don't think that it is necessary.

The strange thing about this is that I used thousands of these bullets in the 1960's and 1970's and they performed much better than those of recent manufacture. When I discovered this core collapse problem, I looked around my

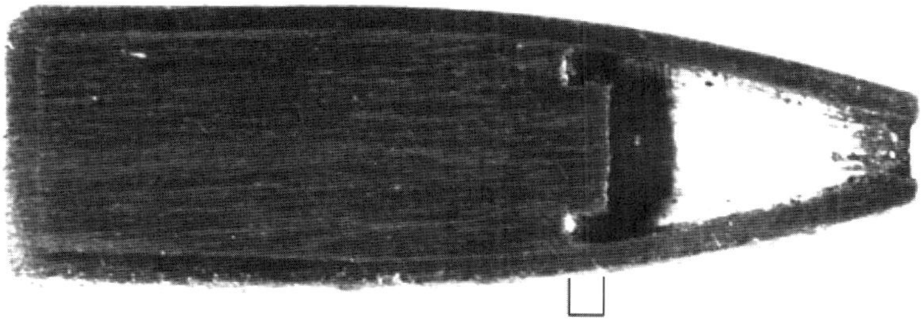

Figure 8-10 - Photograph of a 270 90 grain HP bullet of older manufacture showing the shorter core projection that won't collapse.

shop and found one of the old 270 bullets and sectioned it. Figure 8-10 shows a photograph of this older version of the 270 bullet. You can see that the core projection is much shorter than in the newer bullets (Figure 8-9). In fact, it is short enough that it could not collapse. So, sometime between the late 70's and mid 80's the manufacturer changed the design of this bullet and threw us a curve right in the middle of this research work. However, the reader should realize that this is still an excellent bullet for varmint shooting.

At this point I decided to switch to a 14" twist barrel to see if I could detect an improvement in the 270 accuracy with the 90 grain HP bullet. This would lower the spin-up torque acting on the bullet cores and the 270 should shoot smaller groups. The test results showed that the average group size decreased from 0.804 to 0.505 inches at 100 yards (see Table 11, Chapter 9). However, a lot of this decrease is due to the reduction in dispersion caused by CG asymmetry. Therefore all of the decrease in dispersion can't be attributed to eliminating core striping.

The 6mm 65 gr HP is satisfactory for the 6mm with a 10 inch twist as long as we don't overload it. Unfortunately, these bullets are no longer available after the untimely demise in 1994 of Walter Jankowski, owner of Cook Bullets. So I switched to a 14 inch twist barrel on the 6mm BR bench rest gun and started using the 68 grain match bullets from a different manufacturer. The results of these changes are shown in Chapter 9 which deals with the effects of bullet imbalance.

Effect Of Spin-Up Torque On Accuracy

Every now and then I read an article in a gun magazine about how the rifle rotates as a result of the bullet spin up torque, and how this rotation causes inaccuracy. There are never any facts or data in these articles, just an opinion. Well it is easy to take the spin up torque that we calculated in the early part of this chapter and estimate the amount of angular rotation of the gun. The spin up torque calculated for a 90 grain 270 bullet was 15 inch-pounds. We can scale this up for a 130 grain bullet, and the torque will be 22 inch-pounds. We can estimate the spin moment of inertia of the rifle, and from the moment of inertia and torque we will get an average angular acceleration of 203 rad/sec^2. If we multiply the angular acceleration by the time that the bullet is in the bore (1.3 msec), we get an angular rate of 0.26 rad/sec. We can also calculate the angle that the rifle rotates, and that turns out to be 0.01 degrees. If the CG of the rifle is one inch below the bore centerline, the barrel will be deflected to the left at a rate of 0.26 inches/sec and will translate about 0.013 inches at 100 yards. If the spin torque varies 1% from shot to shot, which is typical of the velocity variation, then the variation in lateral velocity is 0.0026 rad/sec, which will cause a horizontal dispersion of 0.00025 inches at 100 yards. The variation in lateral translation of the barrel is 0.00013 inches. This is a rough engineering estimate of the torque effect, however it can't be off enough to change the conclusion that the torque effect is too small to worry about on a rifle. However, it could be significant on pistols. I didn't check it, so I don't know, but there are so many other problems in pistols, it may not be important.

Summary

The fact that the lead cores in bullets sometimes strip due to the large spin up torques acting on the bullet jacket was demonstrated by measuring the spin rate after muzzle exit using a magnetometer approach. The torque required to cause core stripping was measured with a torque wrench on four test bullets at room temperature (70°F) and at elevated temperatures (250°F). The torque required to strip the cores at room temperature was considerably larger than the torque required at the higher temperatures. Also, harder core material reduces core stripping because it is stronger. These bench tests were in qualitative agreement with the measured spin rate and group size tests.

CHAPTER 9 BULLET IMBALANCE

Bullet imbalance is one of the largest contributors to dispersion, and I have known about it for nearly 30 years. However, the problem was that it has only been in the last several years that I have really understood exactly how it causes dispersion. Also, aside from making or buying perfect bullets, I couldn't find a way to correct the situation. So, let's start out by understanding the problem.

Physical Explanation

Figure 9-1 demonstrates how bullet imbalance causes the bullet to be deflected when it leaves the muzzle. The sketch on the left side of Figure 9-1 shows how the center of gravity (CG), which is offset from the center line or geometric axis, is forced to rotate about the geometric axis. This is an unnatural condition. A spinning projectile will always spin about its principal axis and the principal axis always passes through the projectile CG, if it is free to do so. Consequently, the bullet will start spinning about its principal axis and its CG the instant it exits the muzzle. However, due to the CG offset a tangential velocity component (Vt) was produced while the bullet was in the bore. This tangential velocity component (Vt) will be maintained as a lateral drift velocity (Vd) when the bullet exits the bore. The direction of the lateral drift velocity will be perpendicular to the plane containing both the

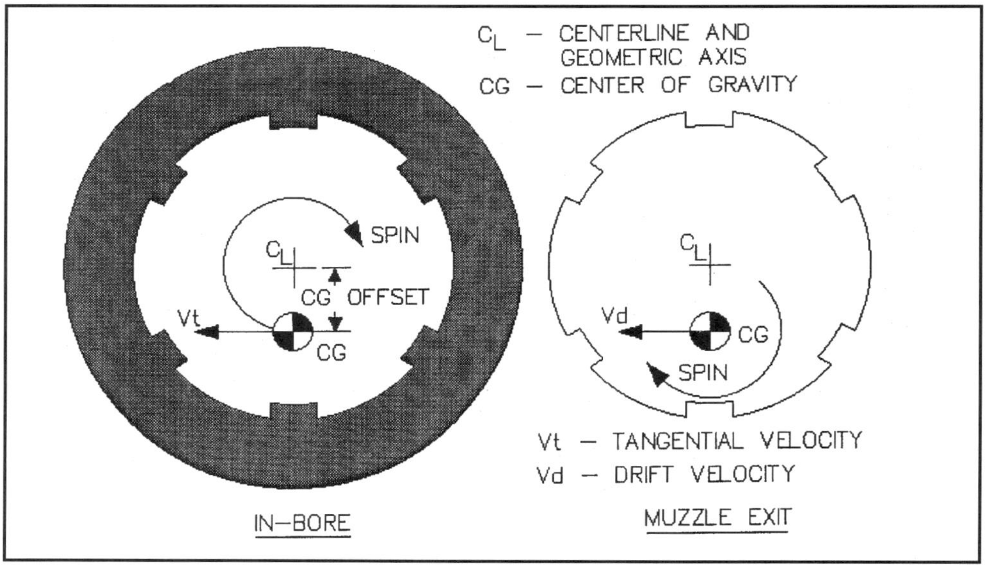

Figure 9-1 - Sketch showing how bullet imbalance causes a lateral drift velocity, which causes deflection of the bullet trajectory as it leaves the muzzle.

geometric and principal axes at the instant of muzzle exit. The distance that the bullet will deflect can be obtained by multiplying the lateral drift velocity by the time of flight. The equation that calculates the amount of bullet deflection at the target is

$$\sigma = 24\,\pi\,(V/t)\,(TOF)\,\delta$$

where

σ = bullet deflection in inches, radius of dispersion or miss distance.

π = pi = 3.14159

V = velocity at the muzzle in fps. Note that V is about 50 to 100 fps less than the instrumental velocity (2900 fps). This results from the muzzle blast continuing to accelerate the bullet after it leaves the bore.

t = twist rate in inches per revolution (10 inches).

TOF = Time of Flight (0.1 second at 100 yards).

δ = CG offset in inches.

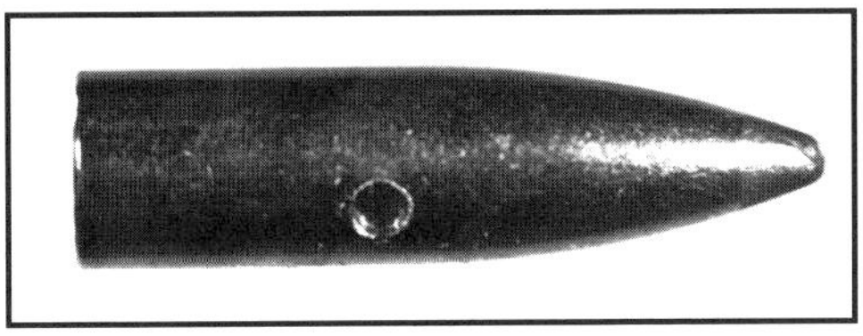

Figure 9-2 - Photograph of a 270 bullet modified by drilling a hole to deliberately produce an exaggerated CG offset of 0.00118 inches.

In the next section we will experimentally determine the radius of dispersion for a CG offset of 0.00118 inches. This is three to four times the maximum CG offset to be expected in a production bullet. This value of CG offset was determined by the diameter and length of the hole drilled in the side of the bullet used in the experiment that follows. Let's calculate the radius of dispersion to be expected at 100 yards from this oversize CG offset.

$$\sigma = 24*3.14159*(2900/10)*0.1*0.00118 = 2.58 \text{ inches}$$

In 1909 Dr. Franklin Mann published a book (Reference 21) with an equation that is equivalent to the one presented here. While his equation was correct and he tested it experimentally his physical reasoning was flawed. However, this was a remarkable book for its time. Now we will experimentally evaluate the effect of CG offset.

Experimental Evaluation

We again turn to the "Olde Engineers Trick" of exaggerating an effect so that it can be easily measured. This time we deliberately unbalance the 90 grain 270 bullets by drilling a hole in the side of the bullet that goes exactly half way through. The hole is placed at the longitudinal CG position. Figure 9-2 shows a picture of a bullet that has been modified to obtain a CG offset of 0.00118 inches. Figure 9-3 is a plot showing the bullet holes from four 3-shot groups fired with thc holc up, right, down, and left at muzzle exit. The square symbols show the center of each group, and the circular sketches near the group show the direction of the hole in the sides of the bullets when they exit the muzzle. If you look at group 1, you can see that the hole in the bullet points up at muzzle exit, which means that the CG of the bullet was below the geometric axis. With a clockwise direction of rotation (right hand twist), the CG in group 1 is translating to the left, which means that the bullet will be deflected to the left, as it was. If you draw a circle with a radius of 2.5 inches, you can see that it passes close to the centers of all four groups. In the

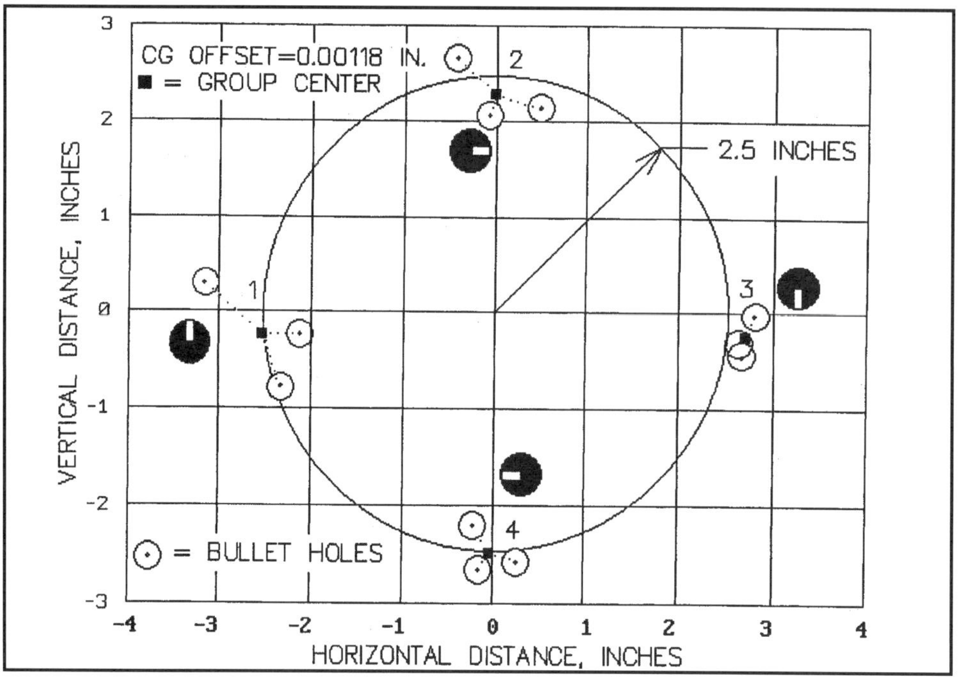

Figure 9-3 - Plot of a target showing four 3-shot groups formed by indexing the bullets in 90 degree increments in roll angle. The bullets had a large CG offset of 0.00118 inches. The experimentally determined radius of dispersion at 100 yards was approximately 2.5 inches.

previous section we calculated a value of 2.58 inches for the radius of dispersion. Roll angle is simply the angle of rotation about the geometric axis (centerline) of the bullet. If you try this test, be sure to remove the extractor and the ejector and have ample headspace between the bolt face and cartridge head. Otherwise you will rotate the cartridge in a random fashion and the results will be a mess. Under ordinary conditions the direction of deflection is completely random, depending on the roll angle orientation of the CG asymmetry. This test confirms our diagnosis of the problem and determines the sensitivity of dispersion to the amount of CG offset. The question now is, how badly balanced are production bullets? Unfortunately, that requires a lot more work, but it can be done.

Measured Bullet Imbalance

There are two ways to measure bullet imbalance — static and dynamic. Static balance is the easiest but least accurate and slowest method.

Figure 9-4 - Photograph of a device that checks the static balance of bullets. The design is based on the principle of the torsional pendulum. See Appendix C for complete description.

Figure 9-4 shows a static balance rig. It is based on the idea of a torsional pendulum where the cradle that holds the bullet is suspended between two lengths of tightly stretched steel wire. As the bullet is rotated, the cradle will rotate if there is a CG offset, and deflect a light beam which produces a light spot on a screen. The motion of this light spot is an indication of the amount of bullet CG offset. The device is balanced by the nut on the screw on the top of the cradle, and the vane hanging down between the two magnets damps the rotational motion. Construction, calibration, and use of this device is described in detail in Appendix C. The results of measuring the CG offset on a box of one hundred 90 grain 270 HP bullets are shown in a bar graph in Figure 9-5. It can be seen that most of the bullets have a CG offset between 0.1 and 0.2 mils, while some of them are unbalanced by about 0.3 mil. This is typical of ordinary commercial bullets. Custom match bullets have about one third this amount of CG offset. Bear in mind that the 90 grain 270 HP is intended for varmint shooting and is certainly accurate enough for that use.

A dynamic balance device is shown in Figure 9-6. In this device, the bullet is spun at 120 revolutions per second (rps) in an air bearing suspended between

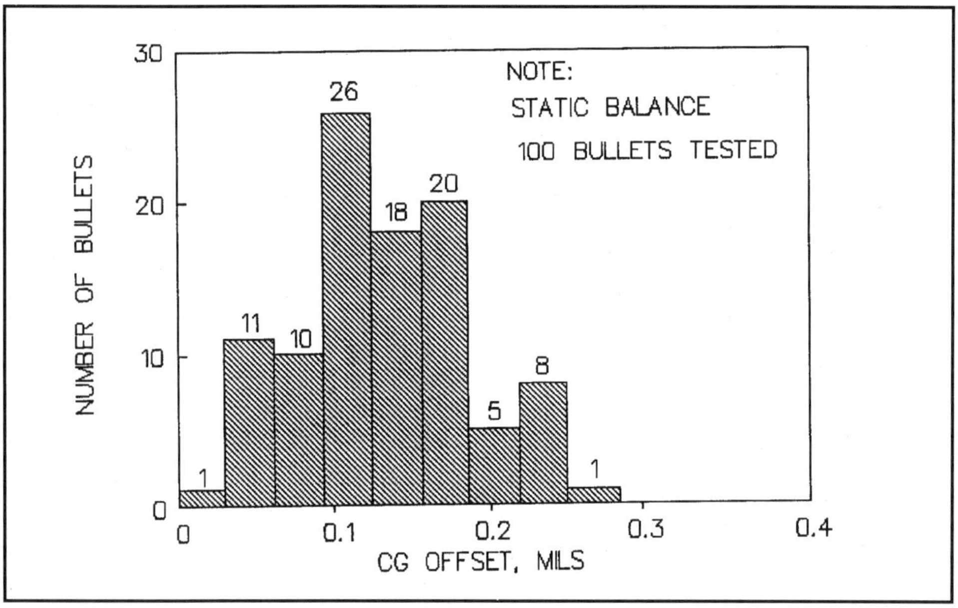

Figure 9-5 - Bar chart showing the results obtained in checking the static balance of a box of 100 caliber 270 bullets.

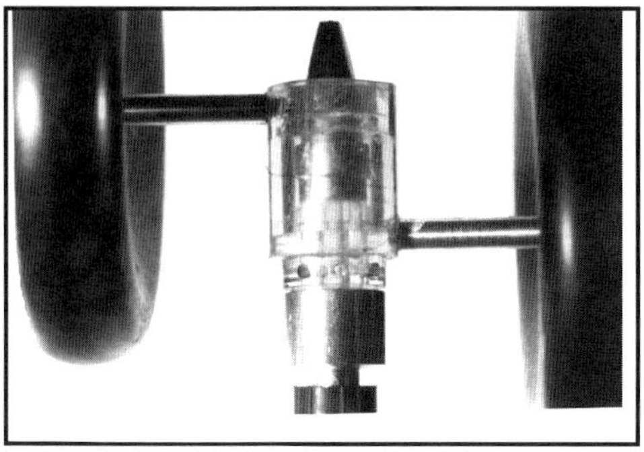

Figure 9-6 - Photograph of a dynamic balance device used to check the balance of bullets. It is based on the principle of the air bearing, where the unbalanced bullet spins inside the plastic cylinder producing an oscillating force on the two earphone diaphragms. This motion produces an oscillating electrical signal proportional to the imbalance. See Appendix C for complete description.

two magnetic microphones that serve as electrical transducers. As the bullet spins, without touching the inside surfaces of the plastic cylinder, the air pressure between the spinning bullet and the walls of the cylindrical cavity force the cylindrical carrier to oscillate. This mechanical oscillation is transmitted to the diaphragms of the two headphones and converted to an electrical signal, which can be observed on an oscilloscope. Construction, calibration, and operation of the dynamic balance device is also described in detail in Appendix C. The results of checking the balance of the same box of one hundred bullets checked by the static balance method is shown in Figure 9-7, and it can be seen that the results are essentially the same. However, the

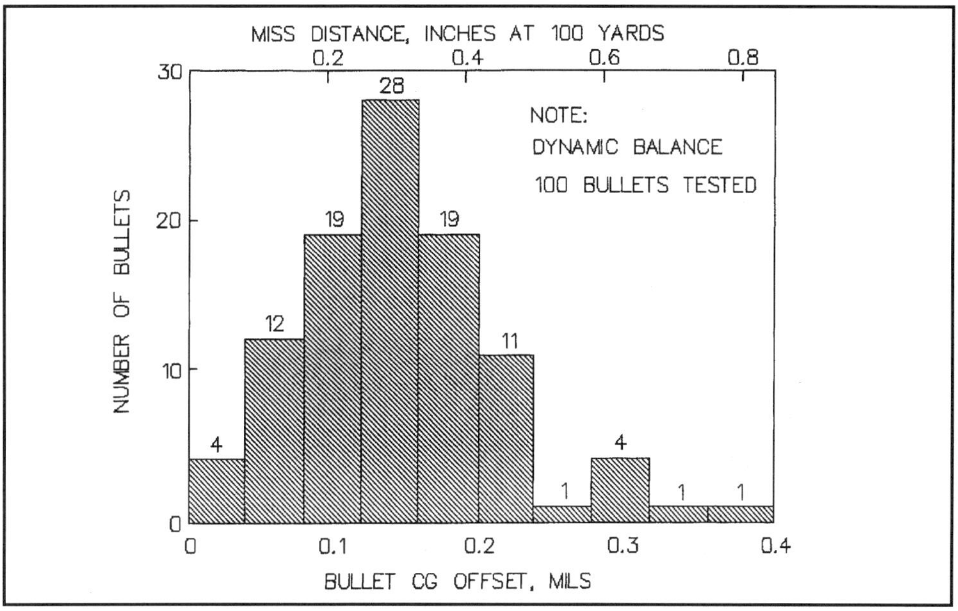

Figure 9-7 - Bar chart showing results of measuring the dynamic imbalance of the same 100 bullets used in the static balance measurement shown in Figure 9-5. Miss distance at 100 yards for a given imbalance is shown on the top scale.

dynamic balance data are smoother and probably more accurate than the data obtained from the static balance device. The dynamic device is much easier to use and is more accurate, but it is much more difficult to make than the static balance device.

At the top of the graph in Figure 9-7 the miss distance (radius of dispersion) in inches is shown for the corresponding CG offset. I developed a computer program that uses a random number generator to pick both the CG offset and roll angle orientation, and "fire" twenty 5-shot groups. I found that the average group size using this computer program with the same 100 bullets in Figure 9-7 would be around 0.7 inches with a 10 inch twist barrel. The maximum computed group size was 1.3 inches and the minimum group size was 0.3 inches. This compares favorably with the last accuracy test fired in Chapter 8, so there is little doubt that bullet imbalance accounts for most of the remaining inaccuracy in the experimental rifle.

I hasten to point out that the measured imbalance on this particular bullet is typical of ordinary production bullets that I have tested. In fact, I have found bullets of other manufacture that were worse. The most likely cause of bullet imbalance is the circumferential variation in jacket wall thickness, that results from deep drawing a flat copper disk to form the jacket. In fact, I am amazed that bullets can be made as accurately as they are in mass production. When you measure bullet jacket thickness run out at the same distance from the base of the bullet, you find circumferential variations consistent with the CG offsets that we measured. The manufacturer states in their brochure that their hunting bullet jackets are held to a maximum of 0.6 mil and their match bullets to 0.3 mil jacket concentricity. Since the jacket of these 90 grain hollow points is about 1/3 of the total weight, the CG offset is about 2/3 of the jacket concentricity. This means that the CG offset will be about 0.4 mil for ordinary bullets and about 0.2 mils for their match bullets. The 0.4 mil imbalance agrees well with the measured data in Figure 9-7. The CG offset is, of course, caused by the fact that lead is much heavier than copper. Some match bullets are held to less than 0.1 mil CG offset. I tested 6mm 68 grain match bullets and got a maximum CG offset of about 0.07 mils. You would never be able to average 0.2 inch five shot groups at 100 yards with a bench rest rifle if the bullets weren't balanced to 0.1 mil or better. Match bullets are shorter than hunting bullets and are made with thinner jackets, which I guess would make the jackets easier to draw with uniform thickness. Unfortunately, the jacket thickness of general purpose bullets has to be kept where it is for reliable expansion characteristics on game animals. Consequently, I doubt that any manufacturer will be able to produce ordinary bullets that are significantly better than they are now, unless somebody comes up with a better way of making bullet jackets. What we need is some way of compensating for bullet imbalance before the bullet leaves the barrel.

Bullet Balance Compensator

The trick to solving the bullet imbalance problem would be to allow the bullet to spin about its centroidal axis before leaving the barrel. The centroidal axis passes through the CG and is parallel to the geometric axis. If this could be done, the barrel would decrease the lateral drift velocity and decrease the effect of bullet imbalance. I tried three approaches to making a compensator and all three attempts failed.

The first approach was to counterbore the muzzle for a distance of three inches. For this to work the radial clearance between the bullet and bore must be small (less than 3mils). The reason for this small clearance is that the corrective effect depends on viscous interaction between the bullet and the barrel. I tried this starting with a 1 mil radial clearance and the groups were enormous. I gradually increased the clearance and at about a 10 mil radial clearance the gun shot about as well as it did before modification. After doing the muzzle blast shadowgraph tests and seeing the small partially burned powder granules traveling along with the bullet, I have doubts about this method ever working. Also, after testing I sectioned the barrel and found that the counterbore was off center. So that may have doomed the test from the start. If anyone wants to try this, I suggest making piloted reamers in 1 mil increments. According to computer calculations it should work, but I may have missed something in the physical model.

Another way to compensate for CG offset would be to allow the barrel to move about the bullet CG before the bullet exits the muzzle. I tried two different approaches and neither one worked. One of them appeared to be trying to work but it drew straight lines of bullet holes as a result of thermal distortion. I tested the barrel on the bench and found that the muzzle warped enough with a modest change in temperature to explain the drift.

While I haven't given up on this problem I decided to go ahead and publish this book because it is a difficult problem that may not be solvable. Meanwhile all you can do is buy the best bullets that you can find. It also should be pointed out that bullet manufacturers are continually trying to improve the quality of their bullets and since this data was taken some time ago the situation may have changed by now.

Bullet Making

I would rather have a root canal operation on a tooth than make my own bullets, but I have been forced to make some special bullets on occasion. There have been a number of articles on custom bullet making and I take issue with some of their recommendations. One of these procedures is lubricating the slugs before they are swaged into cores. Lead wire is cut into slugs that are slightly heavier than the swaged cores. The slugs are then lubricated with a mixture of vaseline and lanolin, although other lubricants have been used. This is usually accomplished by rolling the slugs on a cotton cloth that has been coated with lubricant. Another way is to mix a known amount of lubricant into a known volume of solvent and then dip the slugs into the solution. The solution is drained off and the solvent allowed to evaporate leaving a thin uniform coating of lubricant on the slugs. This would seem to be the preferred method because the coating should be thin and uniform. The lubricated slugs are then swaged into cores in the core swaging die where the excess lead is bled off. The cores are then degreased with a solvent. Methylene chloride is commonly used since the EPA has restricted the use of trichlorethylene and 1,1,1-trichloroethane. The problem is that the solvent is usually used over and over, which results in the concentration of lubricant in the solvent increasing with repeated use. This can result in a thin coating of lubricant being left on the cores. To avoid this problem some bullet makers degrease the cores by passing them through a series of three or four containers of solvent that are frequently replaced so that the last container remains relatively free of solubilized lubricant. This technique requires a lot of solvent but is preferred over repeatedly using the same batch of solvent. The use of lubricant can cause potential problems. If lubricant is left on the cores, then core stripping may occur during bullet spin-up causing dispersion. Also, lubricant could be trapped in the surface of the lead when the slugs are swaged into cores. This would cause a center of gravity offset in the finished bullet. I have never found it necessary to lubricate the slugs prior to swaging them into cores. In fact, I first clean and degrease the slugs by tumbling them in a water solution of detergent (Lemon Joy) before swaging them into cores. I also clean the swaged cores in the same manner just to make sure that they are clean. However, I haven't made the volume of custom bullets that some bench rest shooters make so there may be a need for lubricant in these large volume situations that I'm not aware of.

In this business "Cleanliness is next to Godliness." In fact, match bullets should be made in same type of clean room that is used in the production of electronic chips. All it takes is a very small speck of foreign matter either in or on the lead core or the inside of the jacket to cause the one flyer in one group that causes you to lose a match.

The problem of jacket concentricity is one of the limiting factors at the moment and I have tried to correct jackets with machining with no success. Maybe you could use a boring tool in a super accurate lathe with essentially zero (<100 μinches) spindle runout and improve the jackets, but I doubt it. Lathes this good do exist in large shops but they are expensive and difficult to keep in adjustment. The best match jackets come from a company called J-4 which apparently is connected with Berger Bullets and they are <u>very</u> good.

I think I have already mentioned the fact that I accidentally found a small void in the lead core of a sectioned bullet that would have caused a large CG offset. It was probably caused by a small piece of slag that was in the lead wire. Short of X-raying every core, I don't know how one makes sure the cores are uniform. Of course, this would be prohibitively expensive in production.

Another problem with hollow points is that the top surface of the core may not stay flat and perpendicular during the point swaging operation. The core may also bleed by the edge of the punch in the core swaging operation, causing a flash at the core jacket junction. These problems can cause a CG offset and principal axis misalignment. I have observed this problem on commercial bullets that I sectioned and as you would expect they shot very poorly. Leaving a short (0.06") core projection like we see in Figure 8-10 helps to alleviate this problem by reducing the amount of diameter reduction at the front of the core. Just don't make it too long. This problem makes you wonder what happens to the core when it is swaged into the rifling in the throat during spin up. Does the core stay symmetrical? Nobody knows. You might be able to test this by testing the dynamic balance before and after firing using a very soft recovery. I don't plan on doing it because it would take an enormous amount of effort. It may be that slender nose bullets perform well because there is less length of bullet in contact with the rifling. Some bench rest shooters seem to get superior performance from these bullets with slender noses but I have not had that experience.

There is another problem with hollow point cores that extend too far forward into the ogive nose. When the ogive nose is formed the jacket collapses in short segments and is usually not uniform in thickness. If the core is swaged into this forward portion of the jacket it could produce a CG asymmetry.

Some of the commercial bullet makers are turning out match grade bullets that are pretty good as far as balance is concerned. Commercial bullets have improved a lot in the last 30 years. However, the custom hand made bullets still win practically all of the bench rest matches. I would guess that the difference is in quality of the jackets plus you can discard a bullet that didn't "feel right" during swaging. An ordinary machine doesn't have that capability.

If you decide to make your own bullets be sure to use a slightly rounded heel at the base of the bullet. A sharp corner combined with the rifling lands can produce small fins, which can break off as a result of the muzzle blast. I have seen this in old spark shadowgraphs and recovered bullets and it will cause an asymmetry. As far as I am concerned, making your own bullets is a losing proposition unless you need to try a new idea or you want to do it for the "fun" of it. It would help to be slightly crazy!

Accuracy Test

This is the final accuracy test on the 270 experimental rifle with a 14" twist barrel. You may recall that in the last test in Chapter 7 (Table 8) we had an average group size of 0.804 inches at 100 yards using 90 grain hollow point bullets with a 10" twist barrel. We also estimated with the theory in this chapter that the 270 should average about 0.7 inches at 100 yards, if bullet imbalance was the only error contributing to dispersion. Consequently, we should expect 0.7*10/14 = 0.5 inches average group size with the 14" instead of a 10" twist barrel, with no other rifle errors contributing to dispersion. The results of the test of the 270 with a 14"twist barrel are shown in Table 11.

TABLE 11

270 Winchester Accuracy Test
with 14 inch twist barrel and 90 grain HP bullets

Extreme Spread For Twelve 5-Shot Groups At 100 Yards

Average	Maximum	Minimum
0.505	0.617	0.393

I think this test shows that most of the dispersion left in this experimental rifle is due to bullet imbalance. I also believe that this gun would average around 0.2 inch groups at 100 yards with match grade bullets and a 14" twist barrel. Unfortunately 270 match bullets are unavailable at this time. This concludes our work on the 270 sporter.

CHAPTER 10 EXTERNAL BALLISTICS

External ballistics or flight dynamics is the study of the motion of the bullet after it leaves the muzzle. We have already used the Six Degree Of Freedom (6DOF) trajectory computer program (code) to examine the effect of bullet center of gravity (CG) offset and the effect of muzzle blast on the trajectory of a canted bullet. We will find out how a bullet actually moves in flight and how this effects accuracy. All of the work done in this chapter will be for a right hand twist barrel. Right hand twists are the normal way of rifling barrels but occasionally a few people use left hand twists. In the case of a left hand twist the direction of the coning motion is reversed. We start out with a brief description of the 6DOF computer code.

6DOF Trajectory Code

The 6DOF computer code is an invaluable tool for investigating the detailed motion of a projectile. It does this by solving the three translational equations of motion and the three angular equations of motion. These equations are shown in Appendix D. The three angular equations predict the angular motion about the roll (spin), pitch and yaw axes. The output from the angular motion equations are used in the translational equations to predict where the projectile is going in space. The six equations of motion are solved simultaneously by the computer. In order to use the 6DOF code one must know the initial conditions, several aerodynamic coefficients, and the mass characteristics of the projectile. Unfortunately, these codes are not user friendly and are usually used only by professionals. The biggest problem is finding the aerodynamic coefficients for a particular bullet shape. This requires an extensive library which most people don't have. At any rate, we will be using this code extensively and it is very precise if you know the aerodynamic and mass characteristics.

Gyroscopic Stability

A lot has been written about gyroscopic stability, but most of this material really doesn't show the reader how it effects the motion of the bullet and accuracy. With the 6DOF computer program we can show the angular coning motion of the bullet in detail. Figures 10-1 through 10-4 show the coning motion of a bullet for four gyroscopic stability factors (GS) ranging from 1.13 to 2.98. These figures show the angle of attack in the vertical plane (pitch) on the vertical axis versus the angle of sideslip (yaw) on the horizontal axis where the bullet is launched with an initial angle of attack of about 0.2 degrees in Figures 10-1 to 10-3. A smaller angle of 0.13 degrees was chosen for Figure 10-4. The initial angle of attack of 0.2 degrees was chosen because it is probably typical of the maximum initial angle of attack that would be present in a good rifle with a chamber and throat on the center of the bore. The initial angle of attack is probably considerably less on a good bench rest rifle using cases with turned necks and bullets seated in contact with the lands. The best way to interpret these figures is to imagine that you are viewing the bullet from the rear along the flight path and watching the motion of the nose of the bullet. Of course the bullet is flying along a cork screw trajectory around the average flight path. The effect of the cork screw motion on dispersion is considered later. Notice that the bullet starts out with a high frequency (fast precession) coning motion that damps out fairly quickly and the motion settles down to a lower frequency coning motion (slow precession). The higher the gyroscopic stability the lower the slow precession frequency and the higher the fast precession frequency. However, the higher the gyroscopic stability the faster the slow precession damps. This is desirable because the slow precession coning motion is the most persistent. Note that in the case where GS=1.13 (Figure 10-4) the slow precession grew rapidly from an initial angle of attack of 0.13 degrees to a maximum angle of

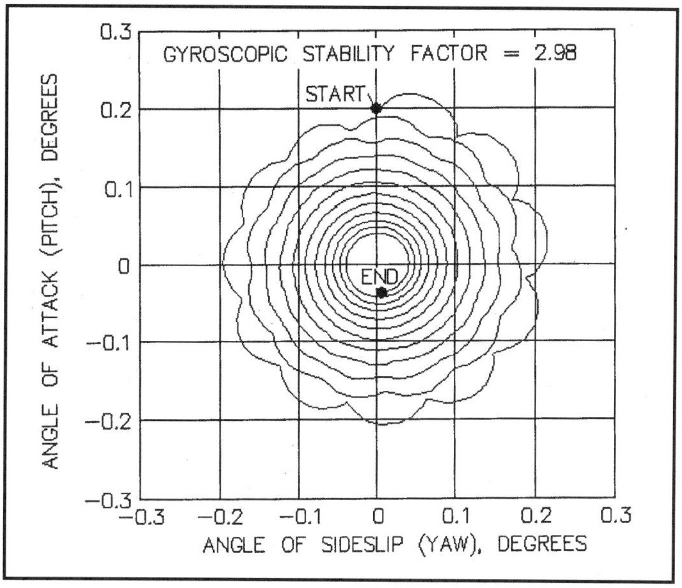

Figure 10-1 - Plot of 6DOF computer flight simulation showing the coning motion of a bullet with a large gyroscopic stability factor (GS) of 2.98. If one were looking along the flight path of the bullet, the motion of the nose of the bullet would appear as the spiral motion seen on the graph. The bullet is launched at the muzzle with an angle of attack of 0.2 degrees and impacts at 200 yards. Notice that there is a high frequency component (fast precession) that quickly damps and a slow component of motion (slow precession) that persists.

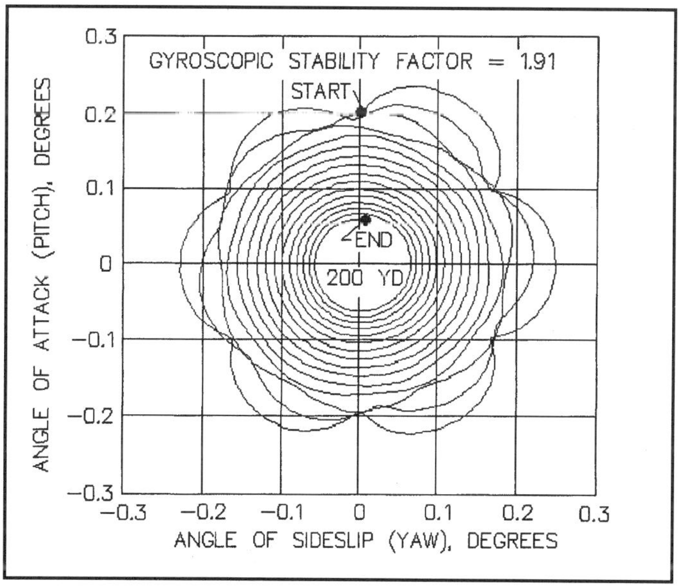

Figure 10-2 - Plot showing the angular motion of a bullet with a GS of 1.91 launched with a 0.2 degree initial angle of attack. Notice that the precession frequencies are slower than those on the previous graph where the GS was 2.98 and that the coning motion takes longer to damp.

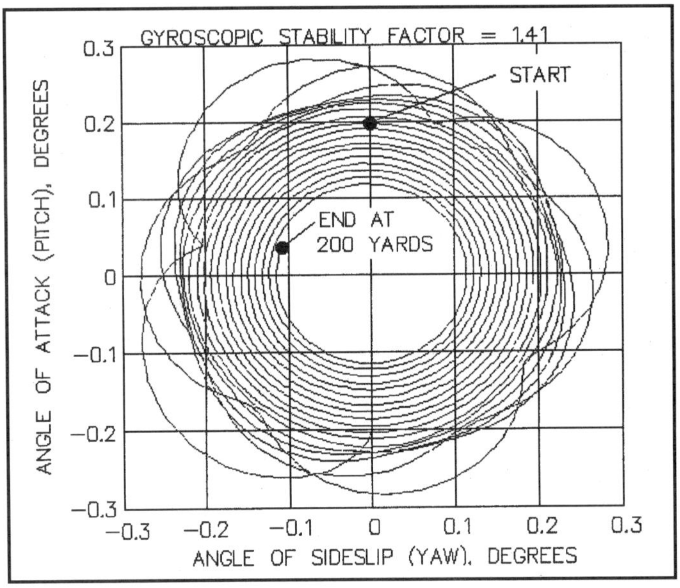

Figure 10-3 - Plot showing the coning motion of a bullet with a GS of 1.41 launched at an angle of attack of 0.2 degrees. This GS is typical of a 6mm 68 grain match bullet.

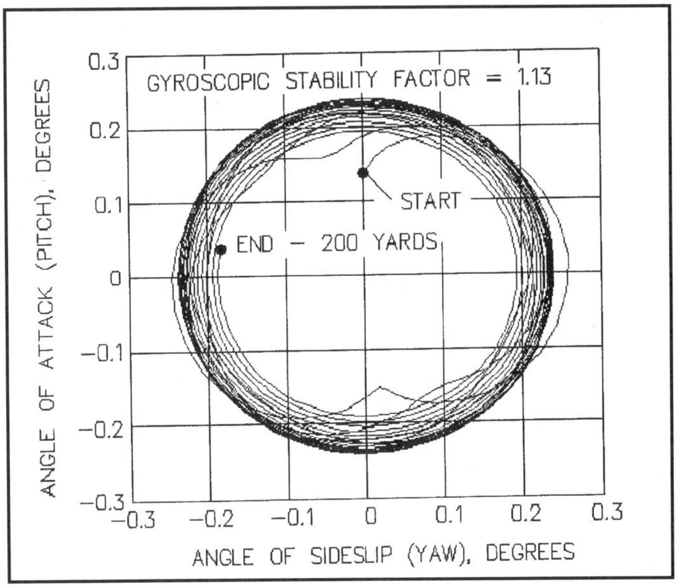

Figure 10-4 - Plot showing the coning motion for a bullet launched at 0.13 degrees angle of attack with a very low GS of 1.13. You can see that the coning motion hardly damps at all over a 200 yard range compared to the previous three figures. Also note that the angle of attack grows rapidly from the initial angle to an angle of 0.25 degrees. This motion is typical of any bullet with a low GS and a normal ogive shape.

attack of 0.25 degrees and has damped down only slightly at a range of 200 yards. These 6DOF computer simulations were run for a 90 grain 270 bullet for twists of 8, 10, 11.6 and 13 inches at an altitude of 5000 feet above sea level just as a demonstration. However, the coning motion is valid for any caliber bullet with an ogive nose and flat base with the same gyroscopic stability factor. At sea level the GS values would be about 16% lower. The typical gyroscopic stability factor for a 6mm 68 grain hollow point bullet is about 1.4 in a 14 inch twist at sea level (Figure 10-3) and would be about 1.6 at 5,000 feet altitude. We will determine the gyroscopic stability factor for a 6mm 68 grain bullet experimentally later in this chapter under wind drift.

Figure 10-5 shows how the maximum coning angle varies with gyroscopic stability. Note that even though the initial angle of attack remains constant at 0.2 degrees the maximum coning angle increases to about 1 degree as gyroscopic stability approaches 1.0. The radius of the corkscrew motion caused by the coning motion also increases rapidly as the GS decreases to 1 causing significant dispersion. You can get away with a low GS (slow twist) at high altitudes and warm temperatures, but a combination of low altitude, high atmospheric pressure and low temperature can shift the data points to the left

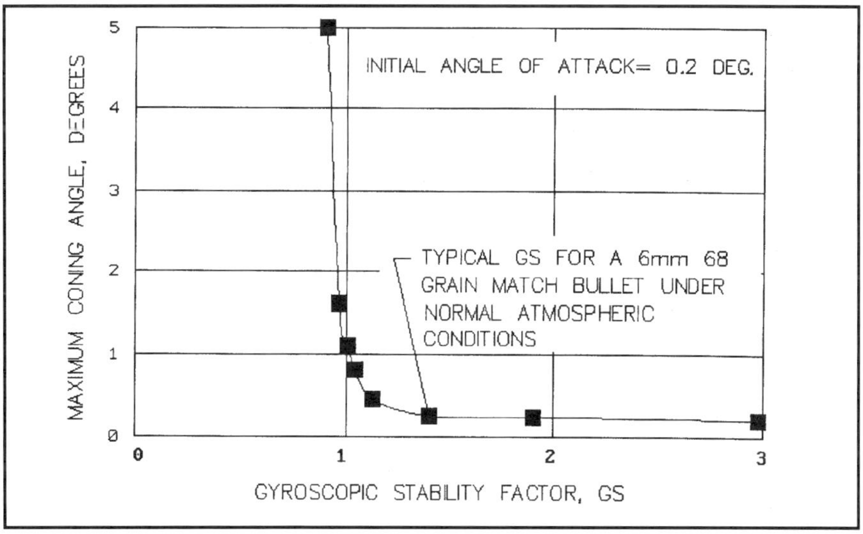

Figure 10-5 - Plot showing how the maximum coning angle varies with gyroscopic stability factor (GS). As GS decreases to one the maximum coning angle and the radius of the corkscrew motion increases very rapidly. Abnormal atmospheric conditions (high pressure, low temperature) will reduce GS by 20% or more. This can cause a normally stable bullet to become violently unstable.

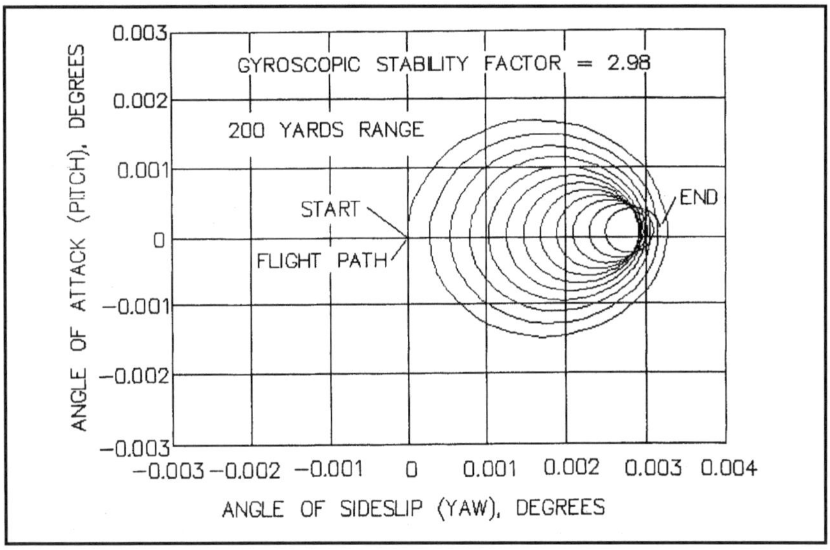

Figure 10-6 - Plot showing the coning motion at a GS of 2.98 where the bullet is launched at a zero angle of attack but gravity drop causes a very small induced angle of attack. Purpose of the plot is to show how the bullet gradually points its nose to the right for a right hand twist as a result of gyroscopic effects known as yaw of repose. Note that the scale is 100 times more sensitive than it was on Figures 10-1 through 10-4 and that the angles are very small.

more than 20%. This would cause a 6mm bullet in a 15 inch twist (GS=1.2) to become gyroscopically unstable, which would result in large dispersion.

Just to show you another feature of the bullet's coning motion a simulation was run that is identical to that shown in Figure 10-1, except that the bullet was launched with no disturbance (Figure 10-6) to show the effect of yaw of repose. The coning motion results from the action of gravity on the bullet causing a small angle of attack as soon as it leaves the muzzle. Note that the scale is 100 times more sensitive than it was in Figures 10-1 through 10-4 and that the angles are very small. You can see in Figure 10-6 that the bullet has a low amplitude coning motion that damps down with the bullet pointed to the right for a right hand twist barrel. This is called yaw of repose and the drift to the right is caused by gyroscopic effects resulting from the downward curvature of the flight path. The yaw of repose angle causes the bullet to drift to the right 0.215 inches in 200 yards. This yaw of repose effect does not cause an accuracy problem because it is consistent from shot to shot. However, it does have an effect on the vertical component of wind drift which we discuss later in this chapter.

It is possible to calculate both the slow and fast precession frequencies and the motion from an analytical theory (Tricyclic Theory) developed to a high degree of sophistication in the 1950's and 1960's by the author and others (References 22 and 23). This is a very useful tool used by the pros to analyze flight dynamics problems. We are going to look at two simple equations that allow one to calculate the two precession frequencies. If the reader is mathematically inclined the equations tell you how the frequencies are effected by various changes. The fast precession (F1) and the slow precession (F2) frequencies are

$$F1 = [(p*Ix)/(2*I) + \{[(p*Ix)/(2*I)]^2 - [M\alpha/I]\}^{1/2}] /(2*\pi), \text{ cps}$$
$$F2 = [(p*Ix)/(2*I) - \{[(p*Ix)/(2*I)]^2 - [M\alpha/I]\}^{1/2}] /(2*\pi), \text{ cps}$$

where

p = spin frequency (spin rate) in radians per second, radians per second = cycles per second * 2π

Ix = spin moment of inertia, slug-ft^2 = pound-ft^2/g

I = lateral moment of inertia, slug-ft^2

Mα = slope of the aerodynamic pitching moment with respect to angle of attack (α)

The moments of inertia can be calculated but they can be more accurately determined by experiment. Now the fast precession frequency (F1) is usually roughly one tenth of the spin frequency (spin rate) and the slow precession frequency (F2) is about one sixth the fast precession frequency for an average length bullet. For instance in Figure 10-2 where the spin frequency is 3600 cycles per second (cps), F1 should be about 360 cps and F2 about 60 cps. Well the actual frequencies in Figure 10-2 for F1 and F2 are about 450 and 75 cps. The reason for this discrepancy is that the Ix/I ratio is larger for a short 90 grain bullet than the normal length 130 grain 270 bullet. In the case of a 90 grain bullet the ratio of F1 to spin frequency should be more like one eighth. A 6mm 68 grain bullet would have similar ratios.

Notice that Mα is always a positive number for a bullet that is aerodynamically unstable but is gyroscopically stable. Note that all normal bullet shapes are aerodynamically unstable and would tumble without being spun at a high

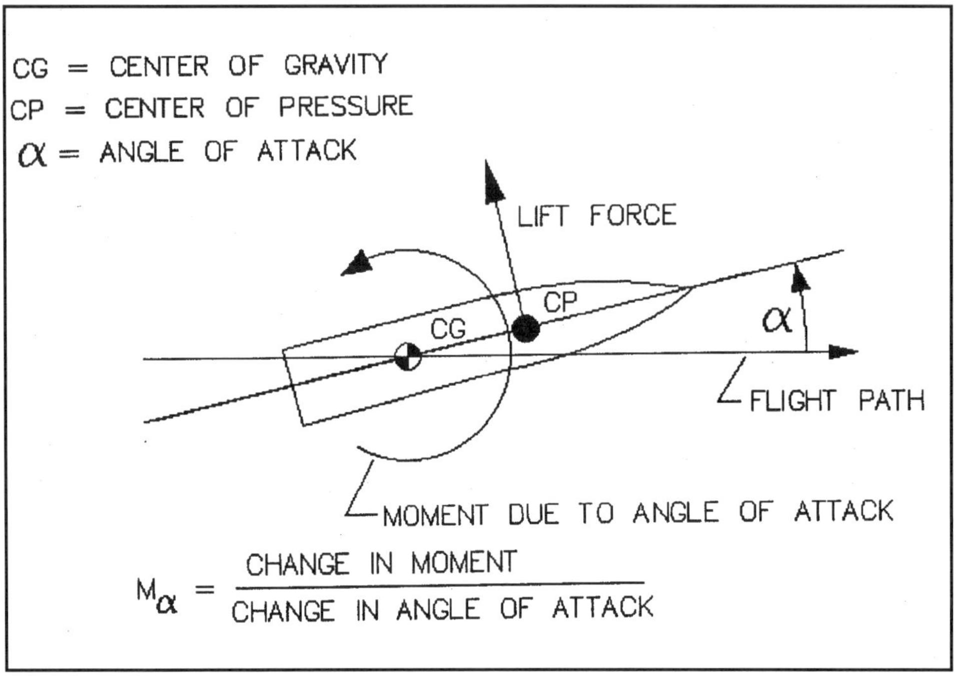

Figure 10-7 - Sketch showing how the lift force acting forward of the center of gravity results in a nose up unstable moment about the CG.

rate. Figure 10-7 shows a sketch of how the aerodynamic pitching moment is developed. When the bullet has an angle of attack (or sideslip or both) with respect to the free stream air flow (or trajectory) a lift force is developed which has a center of pressure ahead of the CG in a normal situation. This causes the bullet to rotate nose up thereby increasing the angle of attack. Consequently, it is an unstable moment. If the gyroscopic stability is large enough it prevents the angle of attack from increasing. The moment on an aerodynamically stable body, such as a rocket with tail fins, has a negative $M\alpha$ or stable pitching moment, because the total lift force acts on the body behind the CG. Now look at the equations for F1 and F2. If $M\alpha/I$ is larger than $[(p*Ix)/(2*I)]^2$ then the square root of a negative number results, which is a "no-no" in mathematics and means that the projectile is gyroscopically unstable. This leads us to define the gyroscopic stability factor (GS) as

$$GS = [(p*Ix)/(2*I)]^2/[M\alpha/I], \quad GS \geq 1$$

where GS must be equal to or greater than 1 for gyroscopic stability. Now, not to worry about trying to calculate this thing, because I will show you a simple way to measure the gyroscopic stability in the section on wind drift.

The reasons for going through all this is first to show you the correct way of calculating gyroscopic stability and how it was derived, but second and more important the equations show you how the gyroscopic stability is affected by the spin rate, moments of inertia and aerodynamics.

For instance, we can show that GS is independent of velocity except for the small effect of Mach number on Mα, because both the square of the spin rate p and Mα are proportional to the square of the velocity. Mach number is simply the velocity V divided by the speed of sound a.

$$M = V/a$$

where the speed of sound is

$$a = 1117*[(°F+460)/519]^{1/2}, \text{ feet/sec} = 1130 \text{ fps @ } 70°F$$

Figure 10-8 shows the effect of Mach number on Mα in aerodynamic coefficient form for the 7.62 mm NATO bullet (Reference 24). You can see that as the bullet slows down on a long range trajectory the Mach number decreases and the moment coefficient (Mα) increases. This means GS decreases at long range and in some cases the effect may be great enough to cause a bullet

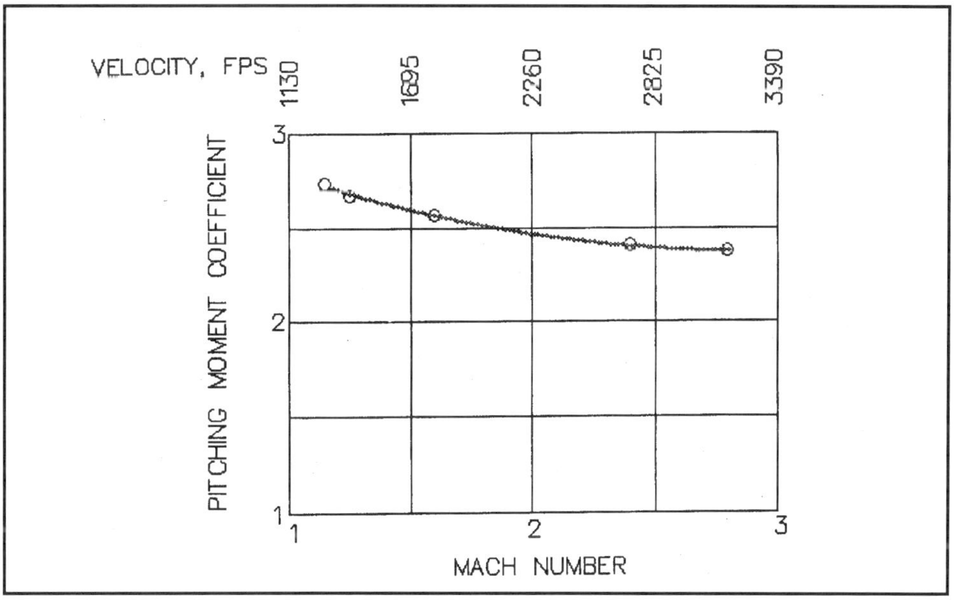

Figure 10-8 - Graph showing how experimental pitching moment coefficient increases with decreasing velocity and Mach number. This is a destabilizing effect.

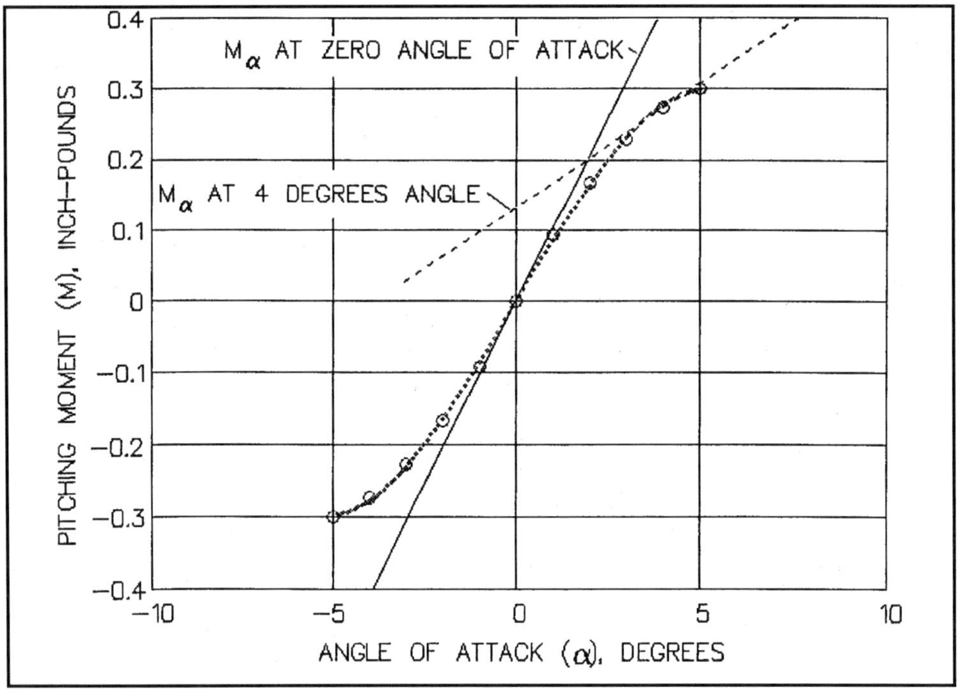

Figure 10-9 - Effect of varying angle of attack on pitching moment slope Mα at 3000 fps. Bullet is more stable at higher angle of attack.

to become unstable. This particular bullet has a GS of 1.3 at 2034 fps (M = 1.8) and a GS of 2.2 at 3164 fps (M = 2.8). This velocity excursion corresponds to a range of roughly 600 yards when the bullet is launched at a muzzle velocity of 3164 fps.

So far we have treated Mα as being a constant with respect to angle of attack (α), but this is only true for small angles. Figure 10-9 shows a typical variation of the pitching moment (M) as a function of angle of attack (α). You can see that the slope of the curve (Mα) decreases as the angle of attack increases. This means that the bullet becomes more stable as the angle of attack increases. This means that a bullet with a GS approaching 1 at long range may be stable at a small coning angle (angle of attack). In this case the nose will travel in a circle at whatever angle it is stable. This is known as a limit cycle and in fact, that is what happens with the 7.62 mm round at long range. So you see, life can get complicated in this business.

Air density has a considerable effect on Mα since it is proportional to density. The higher the density the smaller the GS. The following table shows the effect of altitude on density at an air temperature of 70°F.

TABLE 12

The Affect Of Altitude On Air Density

Altitude (feet)	Density (#/cf)	Density Ratio	Density Factor, σ (1/density ratio)
Sea Level	0.0765	1.0000	1
1,000	0.0743	0.9711	1.0298
2,000	0.0721	0.9428	1.0607
3,000	0.0700	0.9152	1.0927
4,000	0.0679	0.8881	1.1260
5,000	0.0659	0.8617	1.1605
6,000	0.0639	0.8359	1.1963
7,000	0.0620	0.8107	1.2335
8,000	0.0601	0.7860	1.2723
9,000	0.0583	0.7620	1.3123
10,000	0.0565	0.7385	1.3541

Note: #/cf = pounds per cubic foot
Density Ratio = density/(density at sea level)

The density factor is the reciprocal of the density ratio and should be multiplied times the GS. Of course, the GS at 10,000 feet will be 1.35 times that at sea level or 35% greater. What this means is that if you live at sea level your bullets will be more gyroscopically stable at high altitude.

Air temperature also effects the air density. Density is inversely proportional to the ratio of absolute temperature. The absolute temperature (°Rankine) is equal to the temperature in °F added to 459°. For instance the density at 100°F (549°R) is 12% less than it is at 30°F (489°R). This means that the gyroscopic stability will be 12% less at the colder temperature. Atmospheric pressure and humidity also effects air density. Atmospheric pressure can typically vary by 7% between a High and a Low pressure area and have a proportional effect on density at the same temperature. Atmospheric humidity can cause a density change of 2.4% between 0% and 100% humidity. A high humidity decreases density - which is just opposite to what most people would guess. The humidity is usually lowest at low temperatures. So, if you shoot in a high pressure region at a low temperature you could have a gyroscopic stability factor (GS) reduction of 20% or more compared to ideal conditions (low atmospheric pressure and high temperature). If you are using a slow twist barrel (15 inch) where the GS is as low as 1.2 under ideal conditions, you will have an effective GS of less than 1 and the bullets will be unstable.

We should also note that the GS is inversely proportional to the square of the twist rate and inversely proportional to the diameter of the bullet. We should also note that GS is proportional to Ix^2/I. This is why bench rest match bullets are short and light, which maximizes this ratio, and allows the use of a slow twist rate. As we saw in Chapter 9 minimizing the twist rate also minimizes the dispersion error due to CG offset.

So what are the practical effects of GS on accuracy? Well it is important to realize that the bullet is traveling in a corkscrew motion about the trajectory when it is coning. In Figures 10-1 through 10-3 the coning angle at launch is about 0.2 degrees, which is likely to happen. For a 0.2 degree angle of attack the radius of the corkscrew motion will be about 0.009 inches for a GS of 2.98. By the time the bullet reaches 200 yards the angular motion has damped so that the radius of the corkscrew motion is only 0.003 inches. For lower GS's the radius of the corkscrew motion is even smaller. The reason for this behavior can be seen in the equation for the radius of the corkscrew motion.

$$R = q * S * CL\alpha * \alpha * 12 / [(F2)^2 * m]$$

where

R = radius of corkscrew motion, inches

q = dynamic pressure, 1/2 * air density * V^2. Sea level density = 0.00238 slugs/cubic foot (0.0765 pounds/cubic foot)

V = bullet velocity, fps

S = bullet cross section area, ft^2

$CL\alpha$ = slope of lift coefficient, varies from 2.25 for an 8 caliber ogive to 3 for a 6 caliber ogive

α = coning angle, radians. Radians = degrees/57.3

F2 = slow precession frequency, radians/second. Varies from 64 cps at GS=2.98 to 127 cps at a GS of 1.13.

m = bullet weight (pounds) divided by G (G=32.16 ft/sec^2)

This equation agrees very well with 6DOF computer flight simulations. Notice that the radius (R) decreases with increasing slow precession frequency (F2) and since F2 increases with decreasing GS the radius will decrease rapidly with lower GS. In other words, the lower the gyroscopic stability the smaller the radius of the corkscrew motion.

However, this is not the whole story. Note in Figure 10-4 (GS=1.13) that the coning angle jumped up to twice the initial angle of 0.13 degrees and did not damp. As the GS gets smaller and closer to 1.0 the effect increases very rapidly and the projectile never damps. In special low GS cases (GS<1.1) I have had bullets hit the target in a two foot circle. Some shooters are using 6mm barrels with a 15 inch twist instead of the normal 14 inch twist. This will reduce the dispersion due to bullet CG offset by about 7% and may reduce your group size by as much as 5%-6% assuming that CG offset is the major cause of dispersion. It will also reduce the GS from 1.4 to 1.2 under normal conditions and the performance may be erratic under unfavorable atmospheric conditions. Under normal conditions the dispersion caused by the corkscrew motion by itself is too small compared to other error sources to worry about but can become an enormous effect at excessively low gyroscopic stability factors (GS<1.1).

Recall that we tested the effect of muzzle blast pressure on in-bore bullet cant in Chapter 7, which is a much greater effect (0.2 inch radius of dispersion for 0.2 degree bullet cant) and includes the effect of the corkscrew motion. The muzzle blast error was due to the muzzle blast pressure causing a lateral drift velocity and had little to do with the corkscrew motion. However, the test in Chapter 7 was run with a GS of 1.6 and I am sure that the muzzle blast effect would have been greater with a lower GS. So GS gets in the act and effects dispersion even at close range.

Some of the things that can cause an initial angle of attack the instant that the bullet exits the muzzle are in-bore bullet cant, bullet base cant, defect in the bore at the muzzle, and possibly powder combustion products lying in the bottom of the bore. However, bear in mind that the muzzle blast effect is much greater than the corkscrew motion effect that occurs after the bullet leaves the transitional ballistics region at the muzzle. After the bullet leaves the muzzle area there are other disturbing factors that can effect the bullet and introduce an angle of attack and coning motion. A cross wind of 20 mph will cause an initial angle of attack of about 0.5 degrees which will produce a coning motion. At a low GS this initial angle of attack can grow by a factor of two or more. If the cross wind component remains constant from shot to shot there will be no effect on dispersion. However, if you shoot in variable conditions and hold off to correct for wind drift there will be dispersion in addition to the usual wind drift error. The dispersion may be in any direction and not necessarily in the horizontal direction. This error can be larger than those discussed earlier but there is no simple way to evaluate it. There are just too many variables, but a high GS will help. A single tiny rain drop can cause the bullet to rotate to a high angle of attack and result in a significant flyer. Just how bad the flyer will be depends on the size of the drop and where it strikes the bullet. The probability of a bullet hitting a rain drop depends on the density of the rain drops and the length of the trajectory, but it does happen.

The optimum situation is to maintain a GS of 1.4 or greater at a minimum spin rate. If one could move the CG further forward for a given bullet shape this would reduce $M\alpha$ and increase GS. Back in the 1960's I made a 270 bullet with a 150 grain jacket stable at slow twist rates by moving the CG forward. A plastic cylinder was inserted into the base of the jacket and the lead core swaged on top of it (see Figure 10-10). This configuration moved the CG forward with respect to the center of pressure and the bullet was stable in a 16 inch twist barrel. A 150 grain 270 bullet normally requires a 10 inch twist. Consequently, the error due to CG offset was reduced by about 38%. While the accuracy improved in firing tests the accuracy wasn't as good as I had hoped, because the jackets had excessive run out. The 270 bullet weighed about 100 grains so the ballistic coefficient was reduced. This idea might be worth pursuing using 6mm match grade jackets and a twist slower than 14 inches for benchrest competition. For the moment a 6mm 68 grain match bullet at 3200 to 3300 fps with a 14 inch twist is about as good as you can do.

Figure 10-10 - Photograph of a 270 bullet with a light plastic cylinder swaged into the rear of the jacket behind the lead core. This moved the CG forward enough so that the bullet was stable in a 16 twist barrel. This jacket would normally result in a 150 grain bullet, but this bullet weighed 100 grains with the plastic insert.

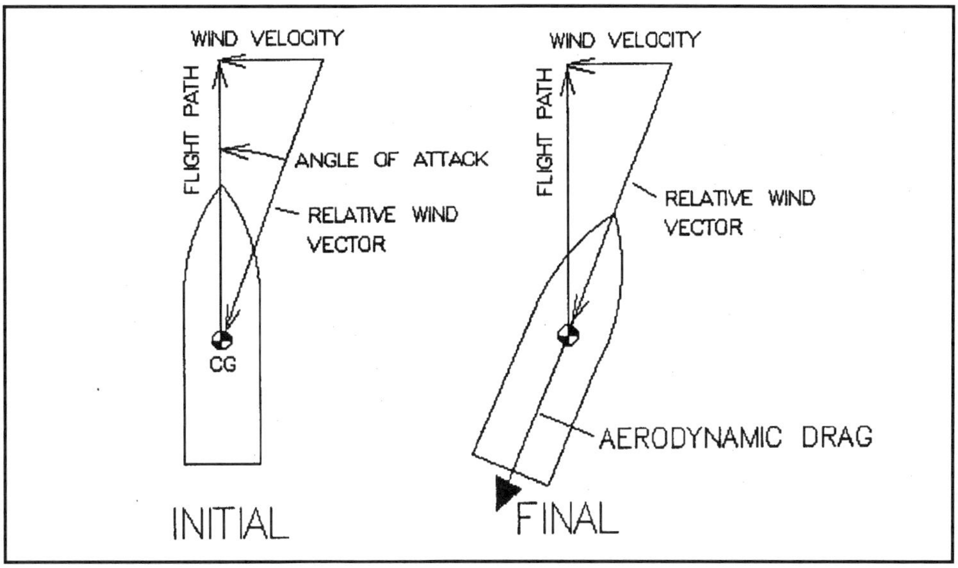

Figure 10-11 - Sketch showing how aerodynamic drag acting along the flight path actually causes wind drift rather than the wind blowing on the side of the bullet.

Wind Drift

Everyone knows that bullets will drift downwind in a horizontal direction but many people don't understand how the drift takes place. Horizontal wind drift is not caused by the wind blowing against the side of the bullet. When a bullet is launched it heads into the wind and the drift is caused by the drag force acting on the bullet, which is canted with respect to the flight path (see Figure 10-11). This sketch shows how the bullet starts out at the muzzle and very quickly aligns itself with the relative wind vector so that the angle of attack approaches zero with respect to the wind vector. In a 20 mph cross wind the centerline of the bullet will be canted at an angle of 0.52 degrees with respect to the flight path. An angle that small or even larger is difficult to detect from distortion of bullet holes. It takes less than one fast precession cycle for the bullet to align itself to the relative wind vector and reduce the angle of attack due to the wind to near zero. When there is no wind the bullet geometric axis lines up with the flight path and the drag force also is lined up with the flight path and there is no wind drift. So, wind drift is not caused by the wind blowing on the side of the bullet as many people think.

A lively discussion recently took place in "Precision Shooting" on how the vertical component of wind drift must be due to Magnus force. Since I am very familiar with Magnus effects (Reference 25) I wrote an article that

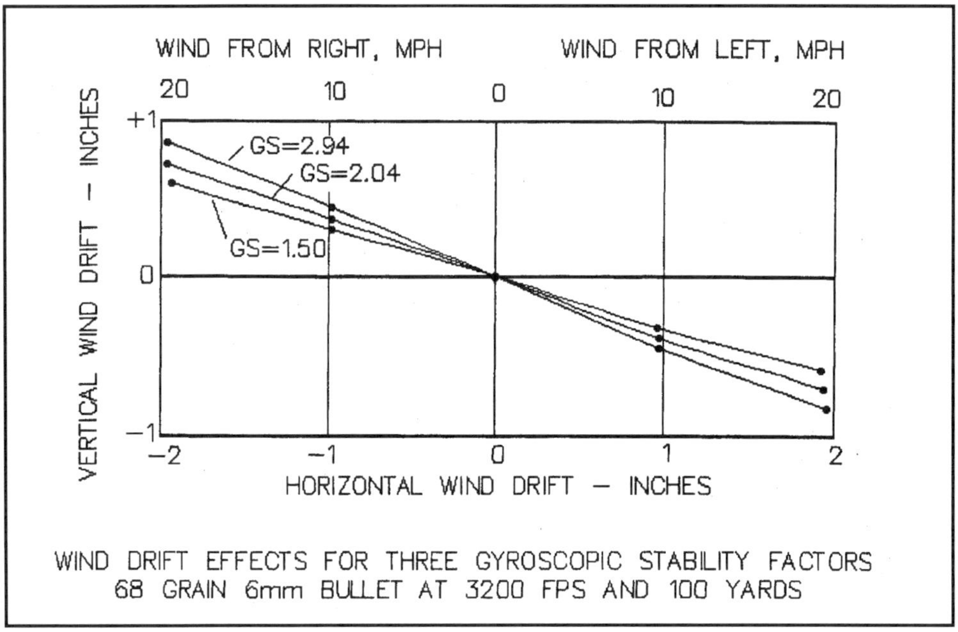

Figure 10-12 - Plot of computer flight simulations of wind drift for three different gyroscopic factors (GS). The drift angle is apparent. The plot is for a right hand twist and the direction of the vertical component would be reversed for a left hand twist.

appeared in the November 1994 issue of "Precision Shooting" explaining that Magnus force acts in the wrong direction and is much too small to cause the observed effect. People also insist on blaming the vertical wind drift effect on rifling marks rotating in a cross wind. Aside from the fact that this would result in just the opposite effect from that which is observed, the rifling marks are buried in a boundary layer that is several times thicker than the depth of the rifling marks. The boundary layer is a thin layer of slowly moving air that forms on the surface of the bullet as a result of air viscosity. This boundary layer tends to blur the effect of small surface irregularities such as rifling marks. Instead of Magnus effects causing the vertical wind drift component it is caused by gyroscopic moments similar to the yaw of repose that we just discussed. Figure 10-12 shows how a 68 grain 6mm bullet will drift in the wind for three different gyroscopic stability factors (GS) as determined from 6DOF computer flight simulation. You can see that the more gyroscopically stable the bullet is, the larger the vertical wind drift component. Table 13 shows the calculated wind drift data in tabular form for both 100 and 200 yards. The data are calculated for a 68 grain 6mm match bullet with a gyroscopic stability factor of 1.5 in a 14 inch twist barrel.

TABLE 13

Calculated Wind Drift Data

Range	Wind Velocity	Horizontal Drift (inches)	Vertical Drift (inches)
100 yards	10 mph	0.962	0.308
	20 mph	1.921	0.578
200 yards	10 mph	4.010	0.608
	20 mph	7.870	1.095

You can see that the horizontal drift component is roughly proportional to the wind velocity and proportional to the square of the range while the vertical component is roughly proportional to both the wind velocity and range.

Figure 10-13 shows a target that resulted from firing a 6BR rail gun at 3200 fps (200 yards) with Berger 68 grain match bullets (14 inch twist) in a wind of varying intensity from the right, and you can see the vertical and horizontal

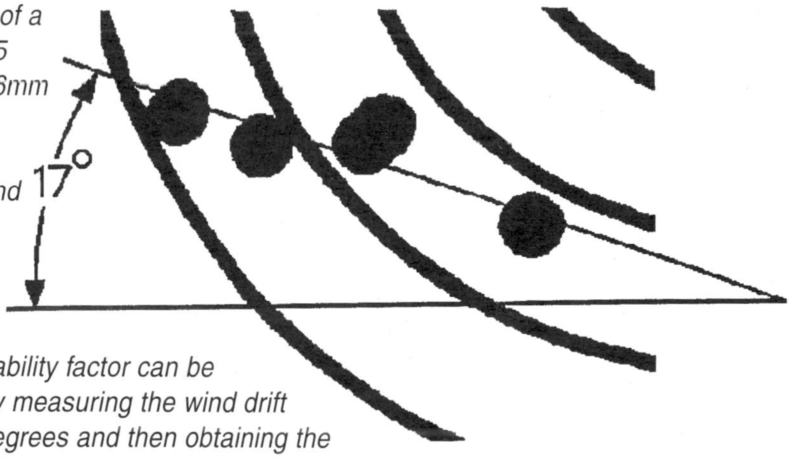

Figure 10-13 - An enlarged plot of a target where 5 shots from a 6mm rail gun were fired at 200 yards. The wind was from the right and varied in intensity. The gyroscopic stability factor can be determined by measuring the wind drift angle of 17 degrees and then obtaining the GS from Figure 10-14. In this case the GS turned out to be 1.60 for this 68 grain bullet with a 14 inch twist at 5000 feet altitude. At sea level the GS would be 1.38 when corrected for the higher density at sea level.

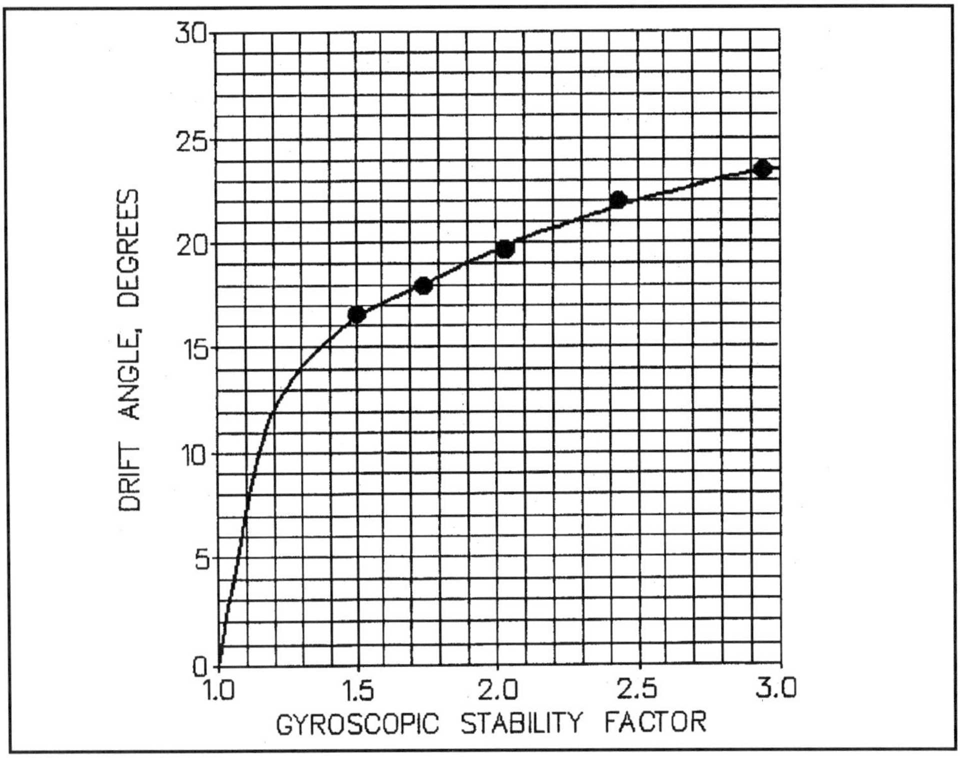

Figure 10-14 - Graph showing how the drift angle varies with gyroscopic stability factor (GS).

component of wind drift. If you measure the angle between a line through the bullet holes and a horizontal line, the angle turns out to be about 17 degrees. If you look at Figure 10-14, which is a plot of the vertical drift angle as a function of GS, you can see that the data in Figure 10-13 (17 degrees) gives you a GS for this particular 68 grain 6mm match bullet of 1.60. If you correct this value for the fact that the test was run at high altitude (5000 feet) using Table 12 (divide by 1.1605) the GS at sea level would be 1.38 which is adequate stability. So, if you are curious about how stable your bullets are you can run this simple test and use Figure 10-14 to find out. It is valid at any range or velocity. You have to fire when the wind is gusting and that way you can get shots at varying wind velocity. The direction of the vertical component of wind drift will reverse with a left hand twist barrel.

Everyone is aware of the fact that bullets drift with a cross range wind. The question is—how much. The most convenient way to determine wind drift is to look it up in tables, such as those in Sierra's reloading manuals. However, if you want to calculate it yourself there is a simple equation that gives good results.

Figure 10-15 - Photograph showing two types of wind indicators (wind flags) commonly used by bench rest shooters. They are both weather vanes which indicate the direction of the wind that have ribbons attached to the tails to indicate wind velocity. The wind indicator on the left has a propellor as another velocity indicator.

$$\text{Horizontal wind drift} = 0.00827 * (v/Va) * (R^2/BC), \text{ at SL}$$

where

v = cross range wind velocity in miles/hour
Va = average bullet velocity over the range in fps
R = range in yards
BC = ballistic coefficient
SL = sea level altitude

I compared this equation with a 6DOF computer simulation for a 150 grain 270 bullet at 400 yards with a 10 mph cross wind and the equation gave 6.96 inches of drift compared to 6.76 inches for the 6DOF calculation (error = 2.8%) at an altitude of 5000 feet. You can use the density correction in Table 12 to correct for altitude. Just divide the result of the drift equation by the density factor. This equation is useful in that it shows that wind drift is directly proportional to wind velocity and the square of the range, while it is inversely proportional to the bullet velocity and ballistic coefficient. This means, of course, that one should maximize muzzle velocity and ballistic coefficient for minimum wind drift.

Wind drift is an important factor in both hunting and target shooting. Having collected most of the North American big game species, including a Grand Slam in sheep, I can appreciate the effect of wind drift in hunting accuracy. The calculations made above were for a 270 Weatherby wildcat cartridge (sharp shoulder) that is my favorite hunting rifle. What I do is memorize the fact that the particular bullet that I am using will drift about 7 inches at 400 yards in a 10 mph wind. Out to 200 yards I don't worry about wind drift (1.75 inches). However, I once had to shoot an elk at 450 yards in a 30 mph wind so I quickly multiplied 3 times 7 in my head and aimed about 2 feet to the right. It hit the elk within a few inches of where I wanted to place the bullet. As a result I had a once in a lifetime trophy that was number 13 in the book. Without that information and a bit of luck, I probably would have missed. The moral of the story is that you hunters had better pay attention to wind drift.

Benchrest shooters and long range target shooters go to a lot of trouble to "dope the wind". A calculation using the drift equation for a 68 grain 6mm Berger match bullet at 3200 fps tells us that the drift at 200 yards will be 3.75 inches for a 10 mph cross wind (or about 1 inch at 100 yards). Well I think you can see that if you are trying to shoot small five shot groups (under 0.5 inches) in a match at 200 yards in windy weather, the wind can be a real problem. Figure 10-15 shows two wind indicators commonly called wind flags that were made by Don Nielson (818-883 5866). The one on the right is essentially a weather vane with a ribbon attached to the tail. The one on the left is the same thing with a propellor on the front of the vane. Typically, three or more wind vanes are placed between the shooter and the target. The trick is to watch the weather vanes for wind direction and to watch the ribbons or the propellor for wind velocity. The problem is to mentally process the six pieces of data and decide when and where to shoot. Some people get very good at doing this instinctively, but it takes a lot of practice. All this mental exercise makes my head hurt, so I built the electronic device shown in Figure 10-16 that does all this mental stuff for you.

The electronic device was originated by Walter Watts (Reference 26) in the late 1960's and he won several big benchrest matches with it. But I don't think it was ever produced commercially. I built it originally to try to minimize wind effects in diagnostic testing before I built the Tunnel Range and it helped. The swiveling vanes are only sensitive to the cross range wind component because they are turned so the plane of the vanes is parallel to the bullet trajectory. The vanes on the electronic gage are mounted on the shaft of a 10 turn 20,000 ohm potentiometer, which forms two arms of a resistance bridge. The bridge is powered by C cell batteries. A two wire cable connects all three of the gage outputs in parallel to an indicator meter on the bench (Figure 10-17). There are three C cells (4.5 v) in the gage nearest the bench, two cells (3 v) in the middle gage and one cell (1.5 v) in the gage near the target. This automatically provides weighting factors of 3/6, 2/6, and 1/6 for the outputs of the three gages. The theory is that the wind drift over the first 1/3 of the trajectory will be 1.5 times that of the second 1/3 of the range and 3 times the drift over the remaining 1/3 of the range. The three gages are placed at 17, 45, and 77 yards from the bench for a 100 yard target. This spacing is approximately the midpoints of the three 33.33 yard intervals in range. According to 6DOF computer calculation this method of correction is quite good. The indicator on the bench (Figure 10-17) is a small box with a microammeter that indicates both plus and minus 50 μamps. When the wind is from the right the needle moves to the right and vice versa. There is an amplifier in the box that allows you to balance the meter and adjust the sensitivity. The sensitivity

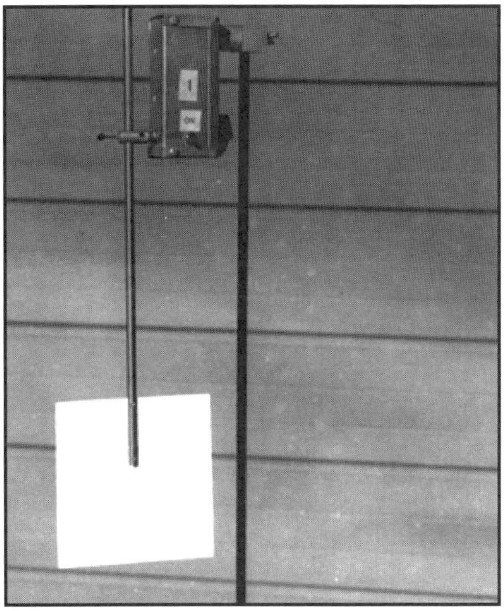

Figure 10-16 - Photograph of an electronic wind indicator which provides an electrical output proportional to the cross wind velocity component. The vane surface is placed parallel to the bullet trajectory so that it is sensitive to the cross range wind component. Three of these gages are used and are connected by a cable to the indicator on the bench shown in the next figure.

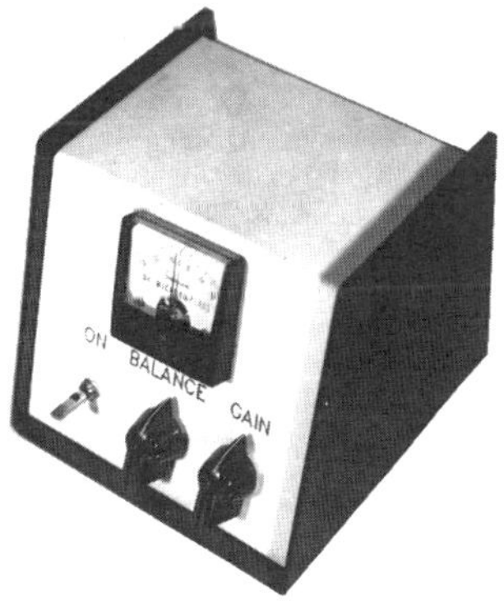

Figure 10-17 - Photograph of the indicator on the bench for the electronic wind gages. When the wind is from the right the meter needle moves to the right and just the opposite for a wind from the left. The goal is to fire when the needle is in the same place.

can also be adjusted by raising or lowering the aluminum arrow shafts that hold the vanes relative to the pivot point. The way you use this device is to observe where the meter needle is pointing most of the time and try to shoot when the needle is at your chosen value. Once in a while you will get caught when the wind suddenly dies off or persists at a higher value and you will have to aim off to correct for the change in drift. However, you can calibrate the effect for a change in conditions on this device by firing on the sighter target. I prefer to use the free recoil method of shooting where the gun is fired by only touching the trigger while the gun sits on the sandbag rests. With the firm hold method I find it difficult to watch both the meter and the scope at the same time. The batteries in this device last a long time (years) and are no problem. The problem with this device is that you have to string 77 yards of cable (or twice that length for a 200 yard match), which is a real nuisance even when it is on a reel. Obviously one could use radio links to get rid of the cable, but this is not a trivial problem and battery life becomes a big problem.

People seem to exaggerate the effect of tail winds or head winds. Intuitively you would think that a head wind would slow the bullet down and make it impact at a lower point. It does, but the effect is much less than most people think. A 6mm 68 grain bullet at 200 yards will strike low by 0.017 inches for a 20 mph headwind or 0.017 inches high for a 20 mph tailwind. The time of flight at 200 yards varies by ±0.4 msec. Now a good 200 yard bench rest group is 0.3 to 0.4 inches in calm weather so I don't think that 0.017 inches for a 20 mph variation in head or tail wind is significant. The effect at 100 yards is less than half the 200 yard effect (±0.007 inches). There is one exception to all this and that is the effect of tail or head winds blowing over obstructions behind or in front of the shooter. On our range I have noted vertical dispersion that I think comes from the downwash created by a tailwind blowing over the roof that covers the benches. This could be minimized by building an electronic gage that is sensitive to vertical wind components. We are also certain that head or tail winds blowing over berms between the bench and target causes vertical dispersion.

We have shown the effect of ballistic coefficient (BC) on wind drift, so it is appropriate to discuss ballistic coefficient.

Ballistic Coefficient

Ballistic coefficient (BC) is simply a numerical value that expresses the ratio of weight to drag for a given projectile. The drag is proportional to the drag coefficient (Cd) and the cross section area which varies as the square of the diameter of the bullet. The ballistic coefficient is

$$BC = 0.0000714 * W / (D^2 * Cd) * \sigma$$

where

W = bullet weight in grains
D = caliber in inches
Cd = drag coefficient
σ = air density factor from Table 12

and the constant takes care of the units involved in the equation. The drag coefficient varies with Mach number (velocity and temperature) and must either be obtained from experiment or theoretical calculation. A sample calculation of BC for a 150 grain 270 bullet goes like this

$$BC = 0.0000714 * 150 / (0.277^2 * 0.30) = 0.465$$

where the bullet is a flat base tangent ogive cylinder with a soft point tip. The drag coefficient was taken from wind tunnel data.

The next thing to do is to find out how to estimate the drag coefficient. One easy way is to get the BC from the manufacturers and rearrange the equation for BC so that you can solve for Cd.

$$Cd = 0.0000714 * W / (D^2 * BC) * \sigma$$

For instance Walt Berger quotes a BC value of 0.276 for his 68 grain hollow point 6mm match bullet at 3000 fps. Consequently,

$$Cd = 0.0000714 * 68 / (0.243^2 * 0.276) * 1 = 0.298$$

This Cd seems about right compared to experimental data in Figure 10-18, which makes the BC seem reasonable. This bullet is a small flat base hollow point that has a very small tip diameter.

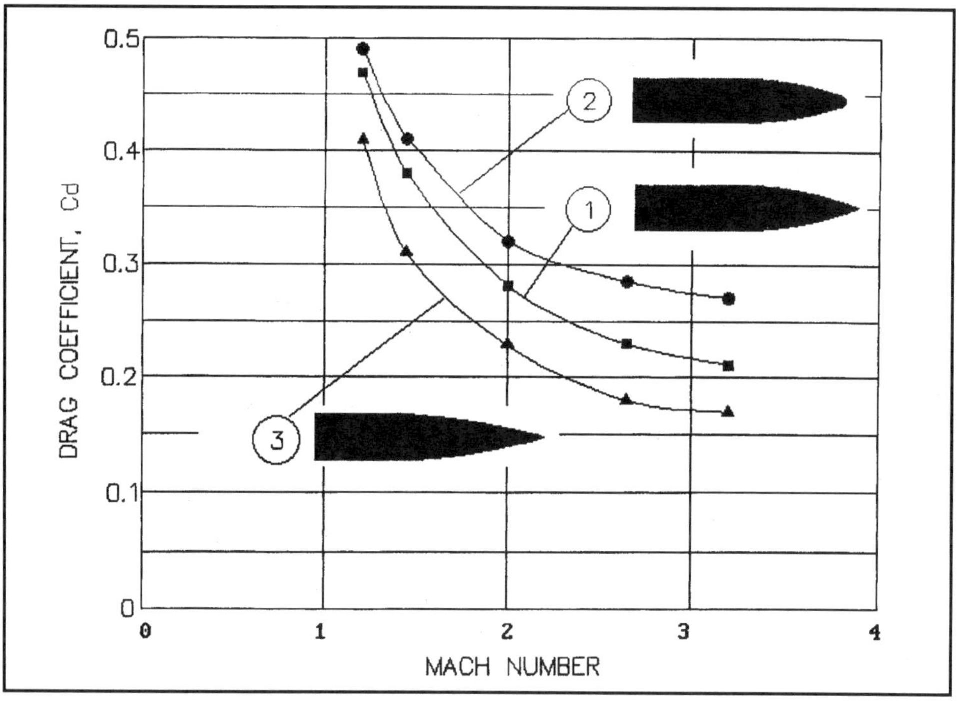

Figure 10-18 - Experimental data showing the effect of nose shape and Mach number on aerodynamic drag at zero angle of attack. Shapes 1 and 2 have tangent ogive noses 2.5 calibers in length, but shape 2 has a blunt nose typical of soft point bullets. Shape 3 has a 3.5 caliber length tangent ogive. Cylindrical afterbody length has only a small effect on drag.

The experimental drag coefficients that have been plotted for three different bullet shapes in Figure 10-18 were taken by O. Walchner in Germany during WWII (1939). I chose to show it to you because I wanted you to know that this kind of data has been around for a long time. Bullet #1 is a 5 caliber tangent ogive cylinder with a nose length of 2.5 calibers and a sharp tip. Bullet #2 has the same shape except the nose tip has been rounded off so that it is similar to soft point commercial bullets. Bullet #3 has a sharper nose that is 3.5 calibers long instead of 2.5 calibers. You can see that the sharper the tip and the longer the ogive the lower the drag. The drag of modern commercial bullets with the sharp-pointed ballistic tips is close to the drag of bullet #1. So, you can see that the BC of the ballistic tip bullets can be as much as 25% greater than the usual soft point.

Bullet #3 is typical of the very low drag bullets that have recently become available, except that they usually have a short boat tail. The boat tail does reduce the drag at low Mach numbers and becomes important at ranges over

500 yards. It has a negligible effect at high Mach numbers and short range (out to 300 yards or so). These long bullets with boat tails are difficult to stabilize and require high twist rates. As a result of the high twist rates and short rifling engraving length, they may be subject to core slippage with soft cores. If this happens the accuracy will be poor. The short engraving length can also increase the tendency for the bullet to tip in the bore.

The effect of bullet afterbody length on drag is very slight. The main effect at high velocity is the shape of the nose as a result of the high pressure acting on the nose. Most of the rest of the drag is caused by the low pressure in the wake acting on the base. The ratio of the head drag to base drag at 3000 fps is 2 or 3 to 1. At lower velocities the base pressure becomes more important relative to the head or form drag and this is why a boat tail becomes more effective at lower velocities or Mach numbers. The skin friction drag developed in the boundary layer is less than 5 percent because of the laminar boundary layer. The effect of rifling marks on drag has been tested and found to be small. The reason is that the rifling depth is only 2 or 3 mils and is buried in the boundary layer. Also, the rifling marks are tangent to the free stream velocity until the bullet slows down. The spin rate slows to some extent but not nearly as fast as the flight velocity.

There is a lot of aerodynamic data available on projectiles but you usually have to have a connection with the military to get access, even though it is unclassified. For instance, Reference 27 published by BRL has aerodynamic data on over 100 projectiles. I think the average shooter is better off either just measuring the BC or accepting the BC published by manufacturers instead of trying to obtain drag coefficient data.

Measuring the BC is really a simple process if you have a chronograph. All you need to do is measure the velocity near the muzzle and the velocity at the range you want to cover. Several shots should be fired and the average velocities should be used in the calculation. Figure 10-19 shows the result of an experimental measurement of velocity at 0, 100, 200, and 270 yards plotted on a semi-log scale. A fundamental equation can be derived that allows calculation of Cd from this data. It is

$$Cd = 0.9221 * W * \ln(Vi/Ve) / (D^2 * R * \sigma)$$

where

W = bullet weight in grains
Vi = initial velocity in fps
Ve = end or final velocity in fps
ln = natural logarithm to base e
D = caliber in inches
R = range in feet
σ = density factor shown in Table 12

So taking the start and final velocities from the figure where R=300 yards, we get

$$Cd = 0.9221 * 180 * \ln(3010/2510) / (0.3085^2 * 900 * 1.1605)$$

$$Cd = 0.303$$

If you look at Figure 10-18 at a Mach number of 2.4, which is the average Mach number over the 300 yard range, you can see that this Cd is about right for a 180 grain Remington bronze point bullet. You can also get BC from

$$BC = 0.00007143 * W / (D^2 * Cd) * \sigma$$

or

$$BC = 0.00007143 * 180 / (0.3085^2 * 0.303) * 1.1605 = 0.517$$

This is a reasonable BC compared to other sources. The function ln is the natural log of the number in parentheses and can be found on most hand calculators. The data were plotted on a natural log scale in Figure 10-19 to show that the equation involving the log function is indeed correct, because the data plot as a straight line. This method gives you a simple way of measuring Cd and BC over any range that you desire. Remember the Cd that you get is nondimensional and depends only on the Mach number. BC depends on Mach number and air density. Many people think that the BC is greatly effected by angle of attack but it isn't in a normal situation. Figure 10-20 shows what is called a drag polar for a typical ogive cylinder bullet with a sharp nose. It shows how the drag coefficient varies with angle of attack. You can see that the drag coefficient increases by only a small amount (less than 1%) at an angle of attack of 1 degree. Back in Chapter 7 we found that the angle of attack at muzzle exit was less than 0.5 degrees on a short 270 bullet.

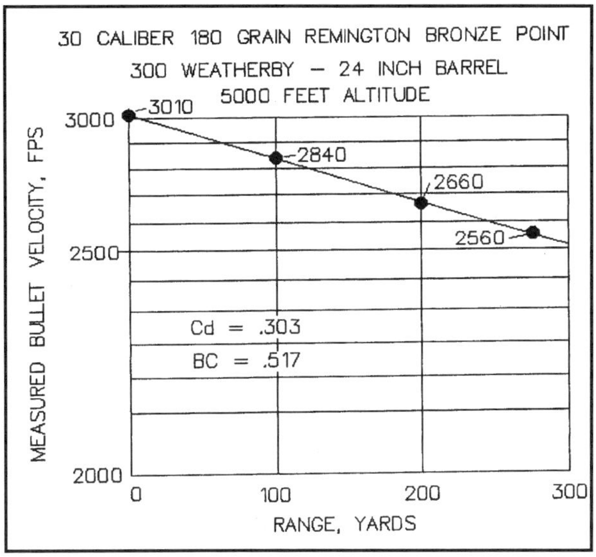

Figure 10-19 - Experimental method of determining drag coefficient and ballistic coefficient. The velocities at 0, 100, 200, and 270 yards were measured on a 180 grain Remington Bronze Point bullet and plotted on a semilog graph to show the logarithmic dependence of velocity on range. The drag coefficient can be calculated from a simple equation shown in the text.

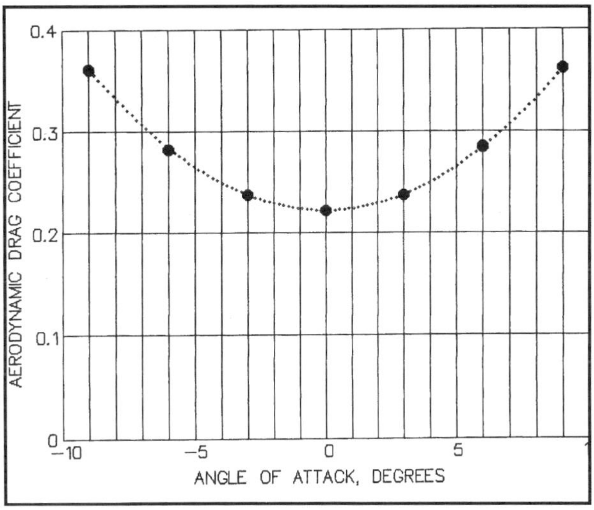

Figure 10-20 - Graph showing how the aerodynamic drag on an ogive cylinder bullet varies with angle of attack. The aerodynamic drag is very insensitive to small changes in angle of attack. From the tests in Chapter 7 we know that the angle of attack at muzzle exit is much less than one degree. A one degree angle increases the drag coefficient and decreases the ballistic coefficient by less than one percent.

The geometry of most bullets simply won't permit large launch angles at the muzzle. If the coning angle is greater than 1 degree it is unstable and BC is the least of your concerns.

I think too much has been made of ballistic coefficient in general. It is important at long range and there you should use a heavy bullet for the caliber at high velocities with a sharp nose and a boat tail. A high BC will minimize wind drift and vertical dispersion due to gravity drop variations. However, it has little effect at 100 or 200 yard ranges where most bench rest matches are fired. In both cases bullet CG asymmetry is more important.

Gravity Drop

One accuracy problem that generally isn't appreciated is the effect of variations in muzzle velocity on gravity drop, which causes vertical dispersion. This effect can be calculated with a simple equation.

$$\delta GD = 385.92 * R^2 * \delta V / (Va\hat{}3), \text{ inches}$$

where

δGD = the difference in gravity drop due to a difference in muzzle velocity

R = range in feet

δV = change in velocity, fps

Va^3 = average velocity over the range cubed

The average velocity over a given range can usually be gotten from a reloading manual. Suppose we have an extreme spread of 30 fps in a 5 shot group at an average velocity of 3000 fps. Then the vertical dispersion due to variation in gravity drop at 100 yards will be 0.039 inches. At 200 yards the vertical dispersion will be about four times that at 100 yards or about 0.16 inches. If you are trying to shoot a 0.2 inch group at 100 yards an error of this size is significant. Another way to estimate this error if you know the total gravity drop at a given range is

$$\delta GD = 2 * GD * \delta V / Va$$

The total gravity drop (GD) can usually be found in some reloading manuals. For instance from the Sierra manual a 70 grain 6mm HP fired at 3100 fps has a total gravity drop at 100 yards of 1.90 inches. For this case the dispersion error for an extreme spread of 30 fps in muzzle velocity is

$$GDe = 2 * 1.90 * 30 / 3000 = 0.038 \text{ inches}$$

which agrees well with the other equation. A 180 grain spitzer boat tail bullet fired at 3200 fps will have a gravity drop error as much as 5 inches at 1000 yards for a δV of 30 fps. The reason for this is the gravity drop at 1000 yards is more than 100 times the drop at 100 yards. The gravity drop goes roughly as the square of the range.

Obviously, the only control we have over this error is to strive for a minimum extreme spread in velocity. About the best that I can do on the average is 15 to 20 fps, which really isn't good enough. As we saw in Chapter 2 filling up the case with powder helps but you may run into excess pressures. Reaming primer flash holes also helps. Some brands of primers seem to do better than others with a particular powder and case. I think that to be competitive in match shooting it is essential to have a chronograph.

Fortunately there is a way to compensate for the velocity variation error. We covered this under Special Bench Rest Gun Problems at the end of Chapter 4 but it may not have been obvious to the reader. If you refer back to Figures 4-39 and 4-41 you can see that the vertical impact point varies as a sine wave with changing muzzle velocity. This is due to barrel vibration and will be different for different guns because the frequency of the vibration will be different. If you shoot at an average velocity that is near a peak and on the negative slope the impact point will be slightly lower for a higher than average velocity and slightly higher for a lower than average velocity. This will compensate for the variation in velocity. These points correspond to 3080 fps and 3330 fps on Figure 4-41. This particular heavy varmint rifle built by custom gunsmith Jim Borden has a Stolle action with a barrel length of 21.5 inches from the front of the action to the muzzle. The problem here is that the optimum low velocity point (3080 fps) requires an excessively light load which might cause increased velocity variation and the optimum higher velocity load point (3330 fps) causes excessive case expansion and may cause core stripping due to the higher chamber pressure. If the barrel were shorter the frequency would be higher and the sine wave would shift to the left. In that case the positive peak could occur at a more optimum load and velocity (3200 fps). The velocity region between the negative and positive peaks is the worst place to shoot because the barrel vibration accentuates the effect of variations in velocity on gravity drop.

Velocity Measurement

I became interested in measuring bullet velocities in 1949 and built my first chronograph in 1950. A chronograph works by counting the number of pulses generated during the time interval it takes for the bullet to trigger the start gate and trigger the stop gate. The pulses are generated by a crystal

Figure 10-21 - Early chronograph built by the author in 1950. It used tube technology and was patterned after the original Potter chronograph. It required 110 volt AC power.

controlled oscillator running at a very precise frequency and the pulses are counted by a series of decade counters. This idea was certainly around in the 1940's and perhaps earlier. I believe the first man to use this idea was named Potter and the first chronographs were named after him. My first chronograph (Figure 10-21) was essentially a Potter chronograph using vacuum tube technology and contact screens. The only problem was that it required 120 vac power. Since you had to plug it in somewhere its usefulness was limited. However, it still works after 48 years and I occasionally use it for other purposes. In 1962 transistors became available and I enlisted the aid of an electrical engineer friend of mine (Harold Bennett) and we built a transistorized version of the original Potter counter (Figure 10-22). It was battery powered

Figure 10-22 - Transistorized battery powered chronograph built by the author in 1962. A contact screen is shown which has aluminum foil cemented to both sides of a piece of cardboard. When a bullet passes through the screen it completes the electrical circuit between the aluminum foil conductors triggering the chronograph.

Figure 10-23 - Photograph of a modern state-of-the-art chronograph (Oehler 35P) with three optical gates. This chronograph uses electronic chip technology for computation and has a built in printer that records all the data. It measures velocity within 2-3 fps at 3000 fps. The optical gates are much more convenient than other types of triggers.

and also used contact screens for gates. One of these contact screens is shown in Figure 10-22. Both of these chronographs had clock speeds of 100 kc which limited the resolution to 0.5% (see Figure 2-18). This is adequate for most purposes but not as good as one should have for diagnostic work. Modern chronographs have a resolution of better that 0.1% with a 6 foot screen spacing. The contact screens are simply a piece of cardboard with aluminum foil glued to both sides. When a bullet passes through the screen it completes an electrical circuit starting the chronograph. These are very accurate gates, but have the disadvantage of not allowing accuracy testing at the same time because they are opaque. Anyhow, I did a lot of work with these instruments in the 1950's through the 1970's.

Modern chronographs, such as the Oehler 35P (Figure 10-23), work the same way, except that they use a faster clock frequency (4 megacycles) and have solid state chip electronics for less battery drain. They also have built in printers and use optical gates that depend on sunlight or electric lights to trigger the counter. In addition they indicate the maximum and minimum velocity, average velocity, extreme spread in velocity, and standard deviation. In the old days we had to compute all this stuff, so things are much faster and easier these days. However, standard deviation is an overkill as far as I am concerned and is meaningless in a small sample (i.e., less than 30 data points). Standard deviation will usually range between 40 and 45 percent of the extreme spread in a 5 shot group. The quantities that are meaningful are average velocity and extreme spread in velocity. The Oehler 35P has three optical screens and it measures the velocity between the first and second screen and between the first and third screen. If the difference between these two velocities is excessive it warns you by printing an asterisk next to the doubtful data. The Oehler chronograph is the best chronograph that I have used and I believe it to be entirely adequate.

The only problem that I have had with the Oehler chronograph is that it is sensitive to elecromagnetic radiation from a radar situated at an airport about a mile away from our range. We found that we could solve the problem by parking a vehicle between the radar and the chronograph. This undoubtedly is a very unusual situation that would rarely be encountered. Glint is another problem with any chronograph that has optical gates. Glint occurs when light is reflected from the ground or some place else onto the bottom of the bullet which can erratically trigger the optical gates. I paint the tube (rail) that the gates are mounted on with flat black paint and put a dark tarp on bare ground to reduce the reflectivity. Glint problems can be difficult to detect, but I have definitely seen it happen when operating on bare sandy soil. Oehler also sells light bulbs that mount on top of the diffusers on the gates for operation in dark conditions. I use these in the Tunnel Range where it is dark and they work very well. However they do require 120 vac power. I prefer a screen spacing of six feet which gives you a measurement precision of 2-3 fps without the length becoming too unwieldy.

The army experimented with several methods of triggering gates including magnetic, capacitance, and optical (called skyscreens). They were all discarded in favor of radar sometime in the 1970's because of the extreme problems with muzzle blast. The orange colored translucent light diffuser mounted above the photodiode (Figure 10-23), that was originated by Oehler, was a big improvement in optical gates.

You will see two terms used in measuring velocity – muzzle velocity and instrumental velocity. Instrumental velocity is the projectile velocity measured at some distance from the muzzle while muzzle velocity is the velocity near the muzzle after leaving the muzzle blast region (10-20 calibers). Muzzle velocity is the instrumental velocity corrected for the loss in velocity between the muzzle and the center of the chronograph gates. If you want to get "picky" about this you can estimate the velocity change between the muzzle and the chronograph velocity from

$$\sigma V = 1.461 * V * D^2 * R * Cd / W$$

where

σV = change in velocity, fps
V = measured velocity, fps
D = caliber, inches
R = distance from muzzle to the center between gates, ft
Cd = drag coefficient
W = bullet weight, grains

For instance, for Cd = 0.3, V = 3200 fps, D = 0.243 inches, W = 68 grains, and R = 8 feet, σV comes out to be 9.7 fps velocity loss. Add that value to the chronograph velocity and you have muzzle velocity.

Since I have already told you more about measuring velocity than anyone ever wanted to know, we consider the effect of rifle cant on accuracy.

Rifle Cant

Rifle cant means rotating the rifle about the bore axis. If the rifle cant angle varies it can have a serious effect on accuracy - particularly horizontal dispersion. I think you can visualize the problem if you consider firing a rifle that is sighted in to hit the aim point so that the sight is adjusted upwards to compensate for the bullet gravity drop. If you were to fire the rifle in the inverted position you would not only have the drop due to the sight compensation but the gravity drop added to it. Consequently, the bullet will strike low by the equivalent of twice the gravity drop. Just to prove this concept I ran an experimental test.

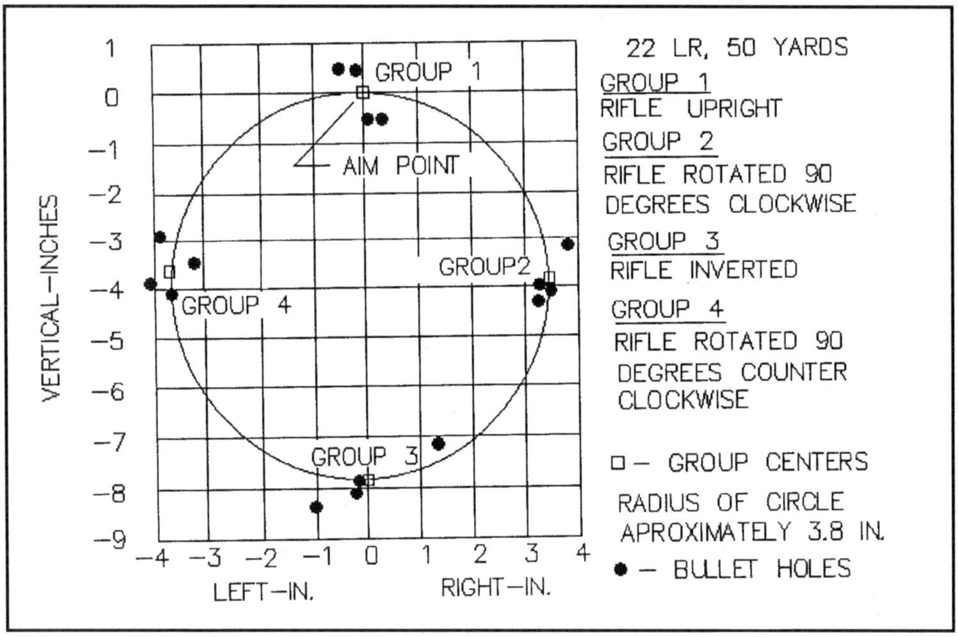

Figure 10-24 - Computer plot of a target showing four groups fired with the rifle vertical, canted 90° clockwise, inverted, and canted 90° counter clockwise.

A 22LR rifle was fired with a cant of 0°, 90° right, 90° left and inverted at a range of 50 yards. Four shot groups were fired at each cant angle and the results are plotted in Figure 10-24. The muzzle velocity and the velocity at the target were measured so that the bullet gravity drop could be accurately computed. You can see that a circle with a radius of 3.8 inches can be drawn through the four groups. Well, I was surprised, because the calculated bullet drop is only 2.7 inches! So where is the extra 1.1 inches coming from? It turns out that the barrel droop due to gravity causes an additional 1.1 inches of drop that must be compensated for by the scope sight. The barrel in the test rifle was a slender cylinder and very flexible. Barrel droop can be calculated very accurately on a cylinder but I won't go into detail because most barrels are much stiffer. So, I felt that this was adequate proof of the concept.

The error can be calculated from two simple equations.

Horizontal error = bullet gravity drop * Sine(cant angle)
Vertical error = bullet gravity drop * (1-Cosine(cant angle))

For small angles (less than 10°) these equations can be simplified to

$$\text{Horizontal error} = GD * \Phi / 57.3$$

where

GD = bullet gravity drop, inches
Φ = rifle cant angle

and the error is in inches. The vertical error is too small to worry about at small angles.

An example is the GD on a 68 grain flat base 6mm bullet at 200 yards is about 7.93 inches with a muzzle velocity of 3200 fps. The horizontal error for a 0.1 degree rifle cant will be about .014 inches or about 0.14 inches for a 1 degree cant. This means that you have to worry about rifle cant in the bench rest game if you can't keep your rifle aligned better than 1 degree. At 1000 yards this effect becomes serious. The bullet gravity drop on a 300 Weatherby with a 200 grain bullet fired at 3000 fps will be about 296 inches at 1000 yards. Therefore, a 1 degree cant will give you a 5.2 inch horizontal error. Since people who win these 1000 yard matches shoot 6 inch groups you really have to be careful about rifle cant.

Hunting is another place where rifle cant can have a significant effect, primarily because you often don't have a good vertical reference in mountain terrain. Suppose you try to hit a big game animal at 300 yards with a 270 130 grain bullet at 2900 fps and you cant the rifle 10 degrees. You will miss your aim point by 3.7 inches. At 500 yards you will get more than three times that amount or about a foot. A ten degree cant angle is fairly easy to have happen in rough country - at least in my experience. So, while not as serious as wind effects, the rifle cant effect is large enough to take seriously at long range.

Figure 10-25 - Anti-cant level device mounted on the barrel of a 36 power Bausch and Lomb target scope just ahead of the eyepiece. It is effective in minimizing rifle cant.

Figure 10-25 shows a bubble level device mounted on a scope that is extremely sensitive to cant angle. It is easy to hold the cant angle to less than 0.1 degree with this device, which is manufactured by DHB Products (phone number 1-(703)836-2648). This is about the only way that I know of minimizing this error at long ranges. At short ranges in target shooting you can sight in the rifle so that the bullet impacts at a distance equal to the gravity drop below the aim point. This amounts to 1.9 inches at 100 yards and about 8 inches at 200 yards for a 68 grain 6mm match bullet at 3200 fps. If you follow this procedure you will effectively eliminate the effect of rifle cant.

Bullet Shape Asymmetries

Bullet tip deformation is one problem in external ballistics that has been explored unsuccessfully in the past. The reason for this is that the effect is so small it is not detectable in experimental tests. However we can estimate the error by running trajectory simulations with the 6DOF computer code.

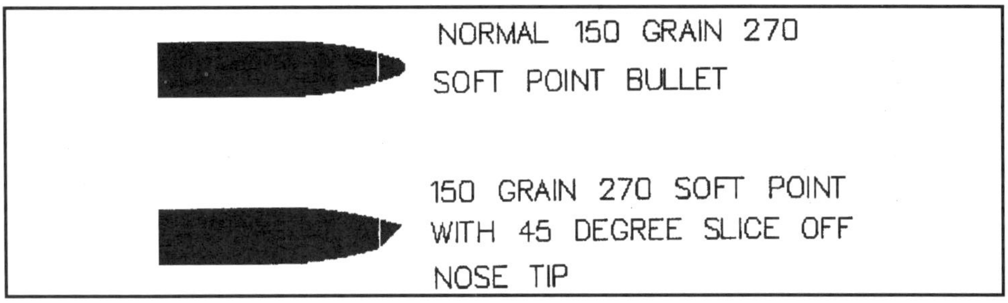

Figure 10-26 - Drawing of a 270 150 grain soft point bullet with a mutilated nose simulated by slicing it off at a 45 degree angle. This type of deformity is not unusual in magazine fed magnum big bore rifles using exposed lead bullet tips.

Figure 10-26 shows a drawing of a 270 150 grain bullet with an exposed lead soft point that has been deformed by cutting off the tip at a 45 degree angle. The results of the trajectory simulations showed that this particular nose tip deformation would cause a radius of dispersion of 0.135 inches at 100 yards. This is not terribly important in a hunting rifle that likely won't group better than an inch at 100 yards. While I have experienced bullet tip deformation of this type and severity in the field, it is unusual to see a bullet deformed this badly. So, at least as far as most hunters are concerned this error is insignificant.

Most bench rest shooters use a match bullet with a small diameter hollow point without an exposed lead tip. This type of bullet would be very hard to deform as badly as the sample case. Some match hollow point bullets out of the box do have a slight angle of the nose flat, that appears to be as much as 5 degrees. If I scale this estimate of 5 degrees angle, the radius of dispersion of a 6mm match bullet would be about 30 to 40 times less than we got on the 270 SP bullet with a deformed nose. This rough estimate indicates a radius of dispersion of 3-5 mils at 100 yards for a 68 grain 6mm match bullet. The smaller the diameter of the hollow point nose flat the smaller the error. This error is essentially insensitive to range, so at long ranges it is completely insignificant.

Bullet deformation is the third mode of motion of a projectile and is called nutation. This nutation mode rotates at the spin velocity of the bullet. The Tricyclic Theory (which means three cycle) includes this nutation mode plus the two modes of precession. Nutation was not included in Figures 10-1 through 10-4 because it would make the graphs confusing. I think you can understand the difficulty in experimentally determining the error contributed by a deformed bullet tip, because it is so small compared to the normal group size. We have already investigated the effect of a canted base in the chapter on muzzle blast (Chapter 7). Small irregularities sometimes occur on the heel (corner of the base) of bullets. I know no way of evaluating such a small irregularity.

Uphill or Downhill

While the error caused by shooting either up a hill or down a hill is unimportant to the target shooter it can be very important to a hunter. The error is easily visualized. If you sight in a rifle at some range—say 300 yards so that the bullet impact is at the aim point then the sight is adjusted to correct for bullet gravity drop. At 300 yards on a high power rifle the gravity drop is about 22 inches. Now if you shoot straight up or straight down the gravitational force will operate along the flight path rather than perpendicular to it. This means that the bullet will impact high relative to the aim point by the amount of the bullet gravity drop (22 inches) regardless of whether you are shooting uphill or downhill. The error can be expressed in equation form as

Up or Downhill Error = GD * Sin(launch angle)
where GD is the gravity drop in inches at a given range.

We can come up with a table for specific angles

Launch Angle(deg)	Sin(launch angle)	Error for GD=22
0	0	0
10	0.174	3.8
20	0.342	7.5
30	0.500	11.0
40	0.642	14.1
50	0.766	16.9
60	0.866	19.1
70	0.940	20.7
80	0.985	21.7
90	1.000	22.0

I think that you can see that if you are shooting either up or down a mountain side that has a typical slope of 40 degrees it is easy to shoot over a medium sized big game animal. You can get into trouble even at shorter ranges. If the rifle is sighted in at 300 yards the midrange trajectory height is about 4.5 inches above the sight line which adds to the GD at 150 yards which is 4 plus 5 inches (9 inches) at a 70 degree angle. I missed a deer once (1950) that was standing on a ledge on a cliff about 150 yards above me. I was shooting almost straight up and the bullet went right over his shoulder. Like most inexperienced people I had assumed that you should aim high on an uphill shot and low on a downhill shot. Not so–you should aim low in both cases.

Bullet Weight Variability

Some bench rest shooters weigh their bullets and separate them into various weight categories. The question is whether or not all this work is worthwhile. I see no way to examine this problem other than to calculate both the internal and external ballistics effects of bullet weight variation. Most production bullets, either hand or machine made, will have a weight variation of about ±0.1 grain about the mean weight on a 68 grain bullet (Figure 10-27). The extreme spread in weight will be between 0.3 and 0.4 grains. Heavier bullets will have a larger variation, but the percentage error will remain about the same.

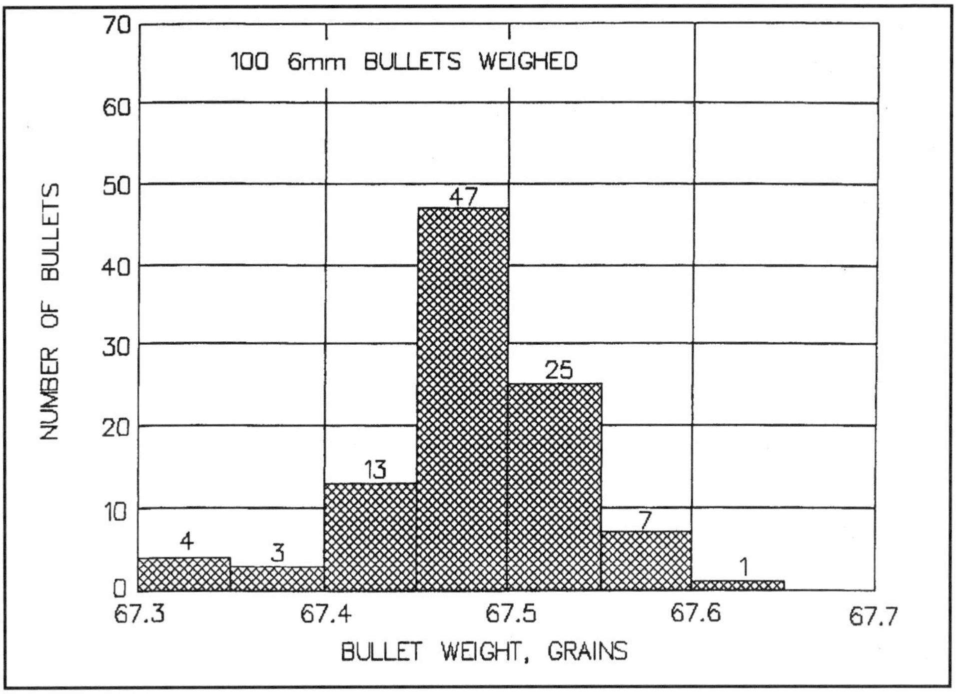

Figure 10-27 - Bar chart showing the weight distribution of a box of one hundred 68 grain match bullets. The extreme spread in weight is 0.35 grains.

So, I made an internal ballistics calculation to obtain the muzzle velocity on a 68 grain and a 68.2 grain bullet to determine the difference in muzzle velocity. It turned out that the heavier bullet was slower by 2.2 fps. I then computed the impact point at 100 yards using the 6DOF trajectory simulation code with the different weight bullet at the different velocities. The two bullets impacted at the same place within 1 mil. In other words, the 0.2 grain difference in weight made no practical difference in the impact point of the bullet. The reason for this is that there are compensating factors involved. For instance, a heavy bullet doesn't result in as low a muzzle velocity as one might think because it causes an increase in pressure over a lighter bullet. This is not a simple proportional or linear problem. If you multiply the muzzle velocity of the 68 grain bullet by 0.2/68 you would get 8.8 fps if the problem was proportional compared to the correct change in velocity of 2.2 fps. Something similar happens in the external ballistics. Even though the heavier bullet starts out slower it doesn't slow down as quickly so that the flight time stays very nearly the same. As a result the bullet drop is very nearly the same.

All of this tells me that sorting bullets according to weight is a waste of time – except for one thing. And that one thing is the possibility of detecting an unbalanced bullet. Sometime ago I was sectioning a bullet when I saw a cavity in the lead core. This was a one in a million discovery that is extremely unlikely to have happen. This small cavity undoubtedly had a flake of some foreign material in it that had been caught and pulled out by the milling cutter. It was probably a piece of slag that would have had a lower density than pure lead and was large enough to have caused a significant CG asymmetry. This bullet was probably lighter than others in the same batch and would have probably been detected by weighing. Unfortunately, I didn't weigh this bullet before sectioning it so we don't know how much the weight would have been reduced. By the way, an air bubble of significant size cannot exist in a lead core because the extreme pressures used in bullet swaging would compress any reasonable size bubble to a microscopic size. However, liquids are essentially incompressible and a drop of oil could cause a bubble in the core. It is important to realize that lead cores are swaged in a constant volume die and the excess lead squirts out of bleed holes. Therefore, the lead cores have the same volume as near as it is possible to make them. However, the lead wire may not have a constant density, which could explain slight variations in bullet weight. Most of the variation in bullet weight (i.e., extreme spread of 0.35 grains) shown in Figure 10-27 does not come from variations in jacket weight and must come from variations in core weight. One hundred J-4 6mm match jackets were weighed using an analytical balance and found to have an extreme spread of only 0.05 grains. If I were going to weigh bullets I would discard the very light ones, because they could have cores containing foreign matter and be unbalanced.

External Ballistics Myths

There are several ideas floating around in the bench rest community that are simply not true. Some of these ideas, which I call myths, are examined.

1) "Increasing spin rate decreases wind drift". It was shown that the vertical component of wind drift is effected by spin rate but the horizontal component is not effected by spin rate.

2) "A good barrel puts the bullet "to sleep" quickly after muzzle exit and therefore the bullet is not effected as much by wind – or those bad conditions". The rate at which the coning motion damps after muzzle exit is only effected by the twist rate, bullet inertia characteristics, and bullet shape (GS). A so called "good barrel," whatever that is, has nothing to do with the bullet's performance in bad conditions.

3) "A bullet goes in and out of stable flight, and if it spends more time in stable flight it will be less effected by the conditions and a good barrel maintains a higher degree of stable flight". A typical bench rest bullet starts out at around 3200 fps (Mach 2.76) and slows down at the rate of roughly 300 fps every 100 yards. At 200 yards the bullet will be moving at 2600 fps (Mach 2.24). This means that the bullet is supersonic through out the first 200 yards of flight and there is no way that the bullet will become unstable if it was stable at the muzzle. Furthermore, if a bullet becomes unstable enough to be effected differently by the wind it will likely completely miss the target. However, at ranges over 600 yards a bullet can slow down enough that it enters the transonic range and be come unstable. As the angle of attack increases it will slow down rap idly and may stabilize again at subsonic velocities. A "good" barrel has no significant effect on the bullet's flight characteristics over a mediocre one.

4) "Cut rifling produces deeper and sharper rifling marks than button rifling and consequently increases the vertical wind drift component". There are several things wrong with this myth. First, the vertical wind drift results from gyroscopic effects and has nothing to do with the rifling marks. Second, you can make rifling grooves at any depth you want with the cut process. I know because I have done it. Third, the boundary layer, where the flow is very slow, is thick enough (more than 3 mils) that the rifling marks are submerged in this layer. Fourth, wind tunnel tests show that rifling marks effect the aerodynamic forces by only a few percent.

5) "I have a load that shoots very small groups at 200 yards but doesn't shoot well at 100 yards". The only way that I can see this happening is for the bullet to be launched with a large disturbance at the muzzle. While this can happen with magnum rifles with excessive muzzle blast pressure, it is very unlikely in the case of a 6BR or a 6PPC bench gun. Most likely, this is a case of poor statistics or a change in conditions.

Summary

There are several important conclusions to be reached in this chapter. If the bullet leaves muzzle at an angle it will immediately start a coning motion which causes a corkscrew trajectory. The magnitude of this angular coning motion depends on the initial angle and the gyroscopic stability factor (GS). For a normal bullet shape the rate at which the coning motion damps with range is only dependent on the GS. The radius of the resulting corkscrew motion and the resulting dispersion is largely determined by the coning angle and the slow precession frequency. It was pointed out that the dispersion caused by the corkscrew motion is usually much smaller than the dispersion due to the effect of muzzle blast pressure on a canted bullet (Chapter 7). This is particularly true for well made bench rest rifles and ammunition. However, at very low GS the dispersion due to the coning motion can become very large. It was shown how the GS can be determined experimentally with a simple test using the vertical and horizontal component of wind drift. We also found out that the GS in normal conditions can be reduced by as much as 20% by a combination of low altitude, high local atmospheric pressure, and low temperature.

The vertical component of wind drift is caused by gyroscopic precession and not by Magnus force as some people think. Methods of measuring winds, gyroscopic stability, and ballistic coefficient were discussed. The effect of nose shape on ballistic coefficient was shown to be significant. The effect of variation of bullet weight and shape asymmetries on dispersion were found to be small. The effect of muzzle velocity variation and rifle cant was found to be significant on bench rest accuracy.

CHAPTER 11 OTHER PROBLEMS

This chapter contains comments on problems that didn't seem to fit in any of the other chapters. Also, problems that are not sufficiently supported by enough experimental data or theoretical analysis to be considered factual are discussed in this chapter. In other words, some of the opinions expressed in this chapter are subject to change when factual data become available.

Bore Fouling and Surface Condition

A lot has been written about bore fouling, which comes from burned powder residue and copper bullet jackets. Most of the bore cleaners either contain ammonium hydroxide or ammonium oleate to dissolve the copper fouling. Examples of aqueous ammonium hydroxide cleaners are Sweet's 7.62 and Parson's Household Ammonia. Long before Sweet's was available I used Parson's Household Ammonia by placing the rifle muzzle down in a coffee can with about two inches of ammonia in it. I then alternated pushing a patch and a brush on a cleaning rod through the bore. I finished the chore by flushing the bore with hot water and drying. Well this method would be very inconvenient to use at the firing range so I normally did this at home. Sweet's is much more convenient to use and works faster than the household ammonia. It also can be used with a patch or brush like any other bore cleaner.

Apparently some barrel makers and gunsmiths have observed bore etching with ammonium hydroxide cleaners. Since I had never noticed this problem I was puzzled by this. My first thought was that it might be caused by residue from the bore cleaner interacting with the powder combustion products at high temperatures (6000°F) to form transient reactive molecular species. At the time I favored the formation of acids. However, after measuring the pH of the residue in the bore after firing and finding it was alkaline (pH=9 to 10) I was forced to conclude that the mixture deposited on the bore was not acid. However, this test does not preclude the formation of transient reactive molecular species at high temperatures. By the way, a pH less than 7 indicates an acid balance and a pH greater than 7 indicates an alkaline balance with a neutral balance occurring at a pH of 7. So, while transient reactive compounds are probably formed at high temperatures, I concluded that an answer to this problem required far more complex testing than I could do. I believe the bore etching problem might be avoided by swabbing the bore with water soaked patches after using aqueous ammonium hydroxide cleaners followed by a cleaner such as Shooters Choice or Hoppe's. I simply clean with Sweet's every 100 rounds or so and follow that with a patch saturated in Shooter's Choice. This works for me.

Shooter's Choice and Flitz both contain ammonium oleate, which is the ammonium salt of oleic acid. They do remove copper fouling, but they are slower than the ammonium hydroxide cleaners. One of my favorite bore cleaners is two parts by volume of Shooter's Choice mixed with one part by volume of Kroil. Kroil is a penetrating oil made by Kano Products and I believe this idea was suggested by Bill Gebhardt, owner of Bald Eagle Precision Machine Company. Liquid Flitz removes copper fouling fairly quickly. It is a precious metal polish that seems to polish the bore surface. After using Flitz copper fouling seems to be slower in forming. You have to use a brush with this stuff and then clean the bore with Shooter's Choice or some other solvent to remove the viscous black gunk that forms.

Chrome-moly barrel steels may react differently to cleaning chemicals than stainless steel barrels. Also barrel steels seem to vary between production runs. This makes it even more difficult to evaluate cleaning methods.

Since everyone has their own idea about how to clean barrels I won't try to tell you how to do it. Instead I will relate to you how some barrel makers and gunsmiths recommend that it be done.

1) Clean every 10 to 20 rounds.

2) Use a bore guide with inserts at the rear that keeps the rod centered to prevent the rod from bowing and rubbing against the rifling. The better bore guides have an O ring that prevents cleaning solution from running back into the action.

3) Use uncoated rods because they are much stiffer and less likely to bend and abrade the lands. Use a jag tip with patches so that they fall off at the muzzle. Pay attention to what you are doing and keep the rod straight when pushing the rod through the bore.

4) Start with 3 patches wet with Shooters Choice–Kroil mixture.

5) Use a brass brush wet with Shooters Choice–Kroil for about 10 complete strokes.

6) Repeat step 4.

7) Run 2 dry patches through the bore and swab out the chamber with a piece of cotton cloth draped on a bore mop.

The first fouler shot will usually have a velocity 50 to 75 fps lower than normal and may not go into the group.

The Outer's Foul Out machine may be the only way to get a bore microscopically clean but it is very slow. The problem is that bore fouling is laid down in alternating layers of combustion products and bullet jacket fouling. As a result the Outer's electronic process stops working when the copper is removed and there is a layer of carbonaceous powder residue remaining. You have to remove the carbon by brushing and then continue the electronic process. However, if you hang in there it will eventually get the bore very clean but it may take a few hours.

There is a form of barrel surface disturbance that has nothing to do with cleaning. If you section a barrel that has been fired a lot, the surface will look like the surface of an alligator bark juniper or the charred surface of a piece of wood (Reference 28). I have seen this in rifle barrels but I am unable to make a legible photograph. The photograph in Reference 28 is very good. A drawing is shown in Figure 11-1 that shows what this surface cracking looks like to the author. This type of surface irregularity is the result of

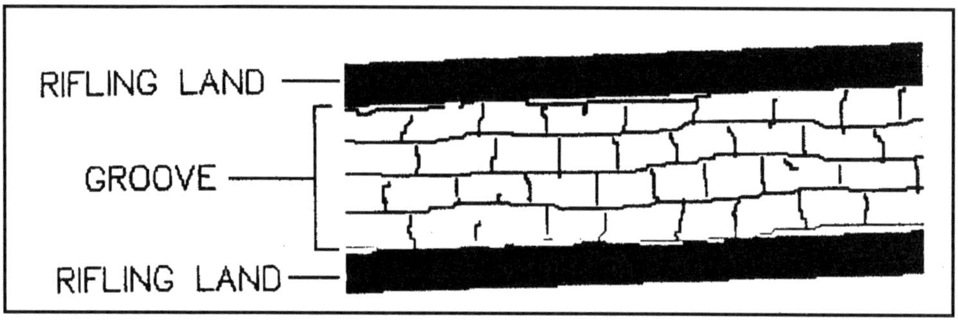

Figure 11-1 - Artist's conception of the interior surface of a rifle barrel showing the effects of surface thermal cracking. This type of surface imperfection is found in barrels that have been fired a large number of times.

thermal compression stress. When a gun is fired the internal bore surface temperatures can reach several hundred degrees and the metal tries to expand. Since the internal surface is restrained by the rest of the barrel it can't expand enough and large compression stresses result. These stresses can be large enough to cause failure of the steel resulting in a minute surface crack. These cracks form in more or less even intervals on the surface resulting in the charred wood appearance. Just what effect this thermal cracking has on accuracy and how soon it appears, I don't know. But I doubt that it is a good effect.

About thirty years ago I tried rifling barrels, because I wanted to experiment with very slow twists and variable twists, which were commercially unavailable. The rifling in these barrels was very rough, because I was very inexperienced. The surprising thing though was that some of these things shot very well. I had used the cut rifling method, which consists of pulling a rifling head with a single cutter through the bore. The cutter is rotated after each pass to give you six grooves, and then the cutter is raised slightly on the next set of passes to cut a deeper groove. This method was automated in production so that a rifling machine automatically indexed the rifling head and adjusted the cutter depth of cut. A lot of custom barrels were made back in those days using the cut process (I still have one), and they were very smooth. Later, practically everyone started using the carbide button swaging process. A button shaped with the desired groove and land shape is pulled or pushed through a bore that is in between the normal land and groove diameters. It is interesting to note that the bore is coated with a light coat of copper before the button is pushed or pulled through the hole. It seems that you can't do it without the thin copper coating which acts as a lubricant. This is used by all the custom barrel makers today that use button rifling, as far as I know. These days some of the large commercial barrel makers use the hammer forging process. A steel billet with a hole in it is hammer forged onto a rifled mandrel. The mandrel is then pulled out of the bore leaving a rifled tube. These three methods were discussed in the March 1993 issue of "Guns And Ammo". The factor that drove the change in rifling methods was cost. However, cheaper is not always better. The question is, which method is likely to produce a better barrel? First of all, I have had no experience with hammer forged barrels, and I haven't seen any data that compares these barrels with the other rifling methods. The forged tubes may not have the reamer marks left on the lands that appear on both cut and button swaged tubes. However, I don't know about the straightness and thermal drift characteristics of the forged barrels. Cut rifling does have sharper corners than button rifling. Whether this is good or bad, I don't know, but I doubt that it makes any difference. Both cut and button rifling have variations in groove diameter. You can feel the tight spots when you lap a bore. Button rifling may have an advantage over cut rifling. The process of swaging work hardens the surface of the metal which should reduce wear. Some people think that a slightly tapered bore is better, but I don't think that it has been proved. In any event, I think most of this argument over which is best is academic, because unless you make your own, you are limited to what you can buy.

The experts all say that a lapped barrel is better. I have to admit that a new barrel that hasn't been lapped shoots better after it has been fired a few hundred times. What probably happens is that the bullet picks up enough carbon and primer grit, which are abrasives, to lap the bore. If you lap it with abrasive before firing you probably save the throat erosion that occurs during the fire lapping process. You can also control the lapping process so that the bore diameter tapers to a smaller diameter at the muzzle. However, barrel lapping is best done by an experienced barrel maker.

A new approach to barrel stress relieving has recently appeared (1995) on the market. It consists of slowly cooling the barrel to liquid nitrogen temperatures and then very slowly allowing it to warm up to above room temperature. This treatment is supposed to improve stress relief and result in a harder steel. Early reports claim reduced bore fouling. However, it is still too early to tell whether the improvement is real.

It is hard to say just how bore fouling effects the bullet, but there is one thing for sure, and that is that a high pressure magnum will lay down a few tenths of a mil of copper in a hundred rounds. This is enough to raise the pressure to dangerous levels when a near maximum load is used. So you should watch the muzzle of these cannons for copper jacket fouling. Meanwhile, the experts are still at odds on the best way to keep a bore clean.

Case Neck Tension

The effect of neck tension on the seated bullet has been discussed in various magazines. None of these articles that I read seemed to have any real data so I decided to try to make some measurements. A load cell was made which measures the force required to seat a bullet in the bullet seating die. The load cell was an aluminum cylinder with two strain gages on it to measure the force. The force was indicated on a milliameter and the peak force required to seat the bullet was recorded. The peak force varied between 30 and 70 pounds. The test was run on a 6mm Remington case with 68 grain match bullets. Sixty rounds were tested and segregated into low (<50) and high (>50) pounds seating force. The rounds were fired in 5 shot groups through an Oheler 35P chronograph in the Tunnel Range with a Heavy Varmint rifle. I could tell no difference at all between the high and low seating force in average muzzle velocity, extreme spread in velocity, or group size. Consequently, I am forced to conclude from the results of this limited test that bullet seating force has no effect on accuracy. However, this test was run on one cartridge and one gun, which is a very limited test and is not necessarily conclusive. It may be that a very light seating force (<10 pounds) may result in uneven bullet seating depth in the lands. This could result in greater dispersion. However, some successful bench rest shooters use very light neck tension and do very well in competition.

Drift Free Bullet

Nearly forty years ago I was heavily involved in launching rockets for research purposes. The first one that we fired turned **into** the wind as soon as it left the launcher and impacted upwind from the intended impact area.

You see the rocket thrust was much greater than the aerodynamic drag which causes wind drift and it over corrected. We came up with a computer code that allowed us to correct the launch angle to compensate for the wind effect. So why not put a small rocket in the base of a bullet that would provide just enough thrust to offset the drag force?

The experience of flying fighter aircraft during WWII that had either 6 or 8 machine guns (50 caliber) on them supported the idea. You see I noticed that while the armour piercing bullets, the incendiary bullets, and the ball ammunition all impacted in the same place, the tracer bullets impacted higher than the others. Since the incendiaries were more visible than the tracers anyhow, I had my crew replace all the tracers with incendiaries. This was a more effective situation because the tracers were wasted. Well I now know why the tracers were so different. It turns out that incendiary bullets have only a 7% reduction in drag compared to the standard round while a tracer has a 40% reduction in drag (Reference 27 and 29). This large reduction in aerodynamic drag is caused primarily by the increase in base pressure resulting from the burning pyrotechnic mixture in the base of the bullet which provides the visible smoke trail. Armed with this information I decided to see if it was possible to eliminate wind drift by placing a small rocket in the base of a bullet that would offset the aerodynamic drag.

Well, the first thing to do with an idea like this is to analyze the problem theoretically and see if it might work. The 6DOF computer calculations said sure enough the wind drift would be eliminated if the drag were zero. The next step was to see if you could squeeze a large enough propellant charge in the rear end of a bullet. The aerodynamic drag on a 6mm bullet is about 1 pound. We know the base pressure drag reduction caused by the hot gasses in the wake can be as high as 40%. So the rocket thrust may only need to be 0.6 pounds.

The better rocket propellants have a specific impulse of about 300 pounds-sec of impulse per pound of propellant. We need about 0.6 pounds thrust for 0.2 seconds (200 yards) and that means we need 2.8 grains of propellant. For these conditions we need a cylindrical grain about 0.210 inches in diameter and about 0.26 inches long. In Chapter 10 I had tried swaging a plastic cylinder into the base of a 270 bullet to increase the stability and it worked fine. With this in mind I tried swaging a cylinder of model rocket propellant into the base of a 270 jacket followed by a lead core. A small hole was drilled into the base of the jacket before the propellant and lead core were inserted. A larger caliber and longer bullet can be made to work at longer ranges, probably to 500 yards.

These prototype bullets were test fired in a strain gage thrust measuring device. This is where the trouble started. It was difficult to ignite the rubber based model rocket propellant without using a black powder electrical squib. When they did ignite with a sufficiently small nozzle hole they occasionally would burn erratically. This unsteady burning is known by rocket engineers as chugging and sounds like a machine gun. So I decided to test fire some of these in a rifle (remotely of course) with a larger than optimum nozzle to see if they would ignite in the barrel. They apparently did not ignite in the barrel because I could not detect any difference in velocity between inert and live rocket rounds at the target.

Faced with the ignition difficulty I decided to give up on this project for the time being because it obviously was going to take a lot more development work than I had anticipated. However, I still think it is a practical idea but I don't plan to work on it in the near future. **I want to warn the reader to take extreme safety precautions if you decide to try this.** In some of the bench tests the jackets exploded but I was protected by a plexiglass box enclosing the experiment plus safety glasses, face shield and padded clothing. Unless you have some experience with explosives don't try it.

Moly Coated Bullets

Recently there has been a lot written in the popular literature about coating bullets with molybdenum disulfide and carnauba wax. The bullets are first coated with molybdenum disulfide (hereafter referred to as moly) and then coated with carnauba wax over the moly. The coating is done by tumbling the bullets in rotary tumbler filled with steel shot. The idea here is that both molybdenum disulfide and carnauba wax are lubricants that should reduce barrel friction and improve performance. I decided to try to test the five claims that have been made for this process.

The first claim is that it is possible to achieve higher muzzle velocities at the same peak chamber pressure due to reduced barrel friction. They also say that at the same powder load a coated bullet will have slightly (3-4%) less muzzle velocity due to the reduction in barrel friction. I tested coated and uncoated 68 grain 6mm bullets for muzzle velocity and chamber pressure. The average muzzle velocity for the uncoated bullets was 3175 compared to 3083 for the coated bullets. The velocity difference was 92 fps or 2.9% which agrees with the claim. The chamber pressure measurements are shown in Figure 11-2 for the uncoated (top photo) and coated bullets (bottom photo). The vertical scale is about 10,000 psi per centimeter. The chamber pressure was about 54,000 psi for the uncoated bullets and about 47,000 psi for the coated bullets. The effect on chamber pressure and velocity was not changed when coated and uncoated bullets were alternated during testing. This indicates that there is no residual effect of the coating and it is just being blown out the barrel. The drop in pressure and velocity is not caused by a reduction in barrel friction as proposed by Norma and others. It is caused by the hot propellant gasses (5640°F) vaporizing the coating resulting in a cooling (about 400°F) of the propellant gasses. The reason that I am certain about this is that if you use a sophisticated internal ballistics code and greatly reduce the barrel friction the pressure drops slightly and the velocity increases slightly. It is physically impossible to get the measured effect on pressure and velocity by reducing barrel friction. Barrel friction has only a small effect on velocity. On the other hand, vaporization of a lubricant takes a lot of energy and a 400°F temperature drop is very likely. Molybdenum disulfide begins to sublime at 842°F and melts at 4802°F. In order to prove this idea I decided to run a test where I simply placed 0.07 grains of moly and 0.07 grains of carnauba wax in the top of the case on the powder. I had found that the difference in weight between the coated and uncoated bullets to be about 0.15 grains. The measured chamber pressure was reduced by about 4500 psi and the average velocity was reduced by 50 fps. This result is similar to the pressure - velocity results that I got when testing the coated and uncoated bullets although the effect was a little less. One could probably fool around with the ratio of moly to wax and achieve identical results to the coated - uncoated bullet test. Anyhow, this test convinced me that molybdenum disulfide cools down the propellant gasses and reduces the pressure. In any event, the loss in chamber pressure has nothing to do with bullet friction. The final step was to increase the load in the 6BR from 27 to 28 grains of

N133 and try to drive the coated bullets at the same pressure level and measure the velocity. The pressure curve was almost identical to the uncoated bullet pressure data shown in the top photo in Figure 11-2 and the velocity was 13 fps higher. Norma claims up to 10 meters/sec (32.3 fps) increase in velocity at the same peak pressure, which is possible. Anyway, I didn't find the increase in velocity performance very encouraging.

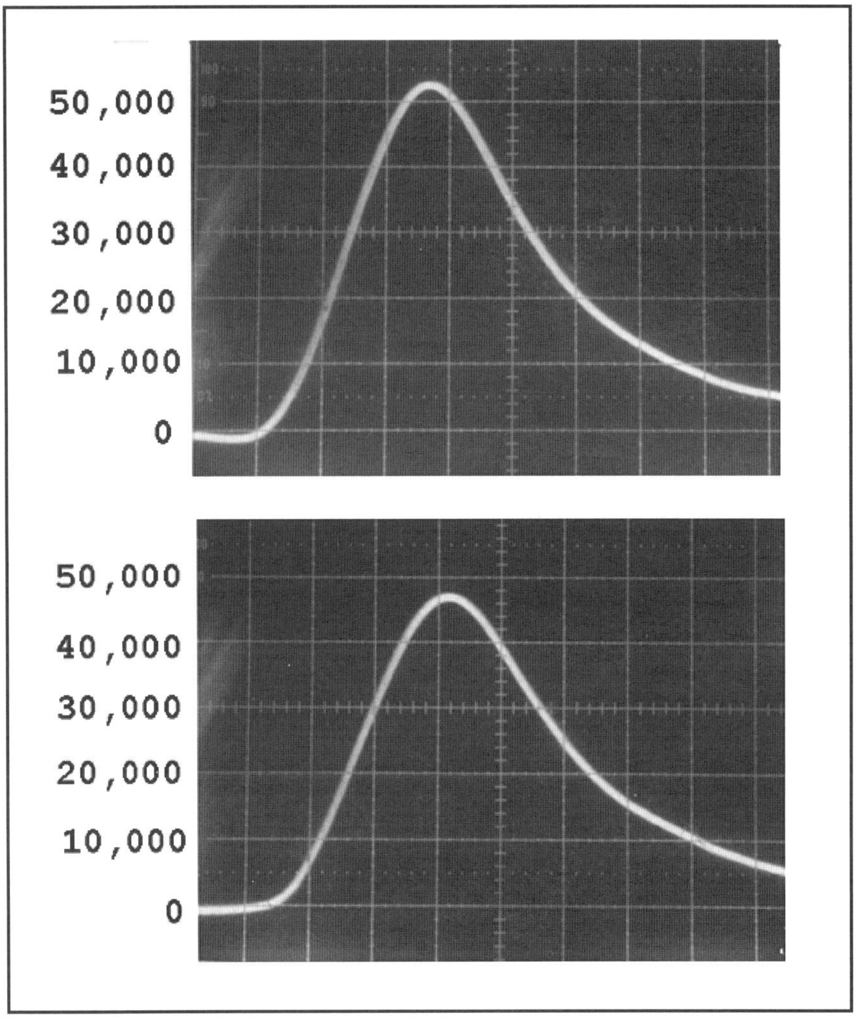

Figure 11-2 - Oscilloscope traces showing chamber pressure measurement with and without molybdenum dissulfide and carnuba wax coating. The top photo shows a peak chamber pressure of about 54,000 psi (muzzle velocity of 3175 fps) for the uncoated bullets and the bottom photo shows a peak chamber pressure of about 47,000 psi (muzzle velocity of 3083 fps) for the coated bullets. N133 powder and 68 grain 6mm match bullets were used.

There have been claims that the coated bullets impact higher on a target at long range (600 yards) implying an increased ballistic coefficient and flatter trajectory. I measured the velocity loss over 100 yards on uncoated and coated bullets. The uncoated bullets lost 325 fps and the coated bullets lost 323 fps over the 100 yards range. The difference of 2 fps is within the limits of measurement accuracy, so I am forced to conclude that there is no difference in ballistic coefficient. Further, there is no reason to expect to see a significant difference in BC because the coating will be gone shortly after the bullet leaves the muzzle if not before muzzle exit. The bullet is hot (600°F) and the aerodynamic boundary layer is hot (750°F), so the wax coating will vanish almost instantly if any is left after going through the barrel. Also, there is no basis for the idea that any lubricant coating can reduce aerodynamic skin friction drag. If it did reduce aerodynamic drag, it could only be a small effect because skin friction drag is only a small part of the total drag on a bullet. Also, the idea that the lubricant somehow reduces the coning angle of attack enough to reduce the aerodynamic drag and increase the ballistic coefficient is not reasonable. This was shown in Chapters 7 and 10 (Figure 10-20). I would think that reports on bullets striking higher on a distant target are due to optimistically increasing the muzzle velocity and chamber pressure or due to firing at a different point on the high frequency barrel vibration curve (Chapter 4).

There are claims of improved accuracy of up to 20%. I did not find this to be the case with my rail gun which averages 0.175 inches at 100 yards. With the same optimum load of 27 grains of H322 and the same lot of bullets coated I got an average group size of 0.179 compared to 0.175 inches for the uncoated bullets (see Figure 4-40). Of course, the muzzle velocity was 92 fps less than it was with uncoated bullets at the same load. When I tried the higher load of 28 grains to obtain the same muzzle velocity as the uncoated bullets, the group size increased to about 0.3 inches. As far as I could see the accuracy was not improved, but this is one gun under one condition. The results could be different with different cartridges or guns. Also I made no attempt to optimize the situation. One thing that did occur to me was that if you don't get a good, even coating of this stuff on the bullet you could make the CG offset worse. Molybdenum disulfide has a density that is about 40% that of lead. So, an uneven coating could make a difference. I believe that I had as good a coating as I could get.

It has been claimed that moly coated bullets have reduced fouling characteristics. I believe this could be true, although I didn't fire enough coated bullets to allow a quantitative estimate of the effect. However, I had the impression that you don't have to clean the bore as often.

Norma claims that barrel life is extended with the coated bullets. I think this may be logical and may be true. After all the pressure and temperature are reduced for the same load which should reduce barrel erosion. However, I don't want to fire tens of thousands of rounds through several barrels to find out. So you can be the judge of that claim.

In summary, I could find no evidence of significantly improved velocity-pressure performance, accuracy, or ballistic coefficient. The coated bullets seemed to require less bore cleaning. The effect on barrel life was not tested. In this limited test I could see no reason for using coated bullets for my purposes. However, you may want to try it because you may get different results.

Shooting Technique

While I don't claim to be a great shot with a rifle, I have made a few observations over a period of 60 years of shooting all kinds of rifles that may be of interest.

I am convinced that the biggest problem in shooting well in hunting big game is getting excited ("buck fever"). By the time I had gotten around to hunting trophy big game I had pretty much gotten over getting excited. I think it just takes experience and it helps to start out hunting varmints. It also helps to get out and shoot at rocks or stumps at unknown ranges. After you shoot at a target of opportunity, pace off the distance. It's easy to get fooled as to the range, particularly in mountainous terrain, and this practice helps. If you can regularly hit a "stump deer" you can probably hit a live deer. Another thing that gets people into trouble is heavy breathing and a pounding pulse. It helps to stop in your tracks and let your breathing and pulse rate slow down. You need to stop and look around often anyhow. I usually move slowly and stop every 50 to 100 feet for as much as a minute. That way you are never winded. This is particularly important for people who live at low altitude and try to hunt at high altitude (say 9000 feet). Another thing that is important is to always try to shoot from a sitting position. Most people practice shooting off a bench at a rifle range. Try to practice shooting in rough country from a sitting position. I don't try to get any closer than 200 yards. That way the animal is undisturbed and I have plenty of time to get into a sitting position and relax. Try it—you'll like it! By the way, don't sight in your hunting rifle by shooting it off a hard sandbag rest. If you do, it very likely will have a different point of impact when you shoot from either an offhand or sitting position. Use something like a rolled up sleeping bag under the forearm. Never use slings wrapped around your arm in the army style because they also affect the point of impact.

Bench rest match shooting is something else and I haven't had a lot of experience at it. There are two main ways to shoot—the firm hold and the free recoil method. While there are variations in the firm hold technique, you usually grasp the pistol grip and hold the butt firmly against the shoulder. We have found that small differences in the firmness of the hold can cause differences in the vertical impact point. Consequently, if you use this approach in match shooting you must be very careful to use a uniform hold while shooting a group. With the free recoil method you aim the rifle by adjusting the front rest and only touch the trigger to fire it. Therefore, you eliminate the problem of holding uniformly and also you can watch all the wind flags constantly. The disadvantage is that the free recoil method is slower than the firm hold method. The only thing to do is to try both methods and see which one works best for you.

One thing that may be a problem is the hard sandbags used in bench rest shooting. I know for certain that guns with heavy recoil don't shoot well off a standard bench rest setup. Just where this starts being a contributor to dispersion is difficult to say but I am suspicious of these hard bags even in the case of 6mm bench guns. Unfortunately the bench rest rules specify the use of sandbags.

Statistical Error

I often see magazine articles where someone makes extravagant claims and presents as few as two 3-shot groups as experimental proof. Well, this is not enough data to prove anything. A series of 3-shot groups will average about half to two thirds the dispersion of a a series of 5-shot groups. Also, it takes at least five 5-shot groups to begin to have any confidence in the data. The average group size data reported in this book are based on a minimum of eight 5-shot groups. However, it usually is not necessary to fire more than ten 5-shot groups to have reliable data.

A Final Word

I have tried to present facts about rifle accuracy as much as possible. Obviously, there are still problems that remain to be solved but I believe we know a lot more about rifle accuracy than we did when I started this research. I am still working on some of the unsolved problems and may eventually write a second volume. The computer programs used in this work will not be available to the general reader because they are not user friendly and I don't have time to thoroughly document them. Also, I do not build guns for people. I hope Remington is not offended by some of my analyses of the Remington 721 rifle. As a matter of fact, the Remington 700 rifle is my favorite sporter and I have several. My objective was simply to improve it. Anyhow, I hope you have found this interesting reading. Good Shooting!

The reader is warned one last time that some of the experiments performed for the book can be dangerous and should not be duplicated by people who do not have an extensive background in test techniques and in explosive technology.

A

APPENDIX-A ACCELEROMETER DESIGN

Early in this work it was discovered that, although a sporter barrel vibrates in many modes, the most important was the third mode at a frequency of about 1.25 kHz. The higher modes simply don't contribute much to dispersion, because even though relatively large accelerations are present at these high frequencies, the displacement caused by these modes is very small. Therefore, the task was to design an accelerometer that would predominantly respond faithfully to this third mode. There are several types of accelerometer designs that are commonly used. Three types were tried in this investigation.

1) piezoelectric crystal - Many commercial accelerometers are of the piezoelectric crystal type. They are sensitive, having a high electrical output, and respond well over a large frequency range. However, they are sensitive to shock in all directions, and tend to vibrate at their natural frequency, which can be 30 kc and higher. I made several of these, using a Radio Shack crystal microphone, and it worked very well at low accelerations when tested on the bench where there is little or no shock excitation. However, when tested on the muzzle of the rifle, where there are a lot of shock waves running through the barrel, the data were obscured by a lot of high frequency oscillation. In fact, it appeared that the instrument was driven into saturation at the high frequencies. When the output was run through a low pass electronic filter, the lower frequency data

that is of interest began to appear. However, when the output of the accelerometer is partially saturated, the low frequency data are likely to be badly distorted. Attempts to damp the accelerometer, in an effort to attenuate the motion of the sensing crystal, were completely unsuccessful. Perhaps one could do better with a commercial accelerometer of this type, but I believe one would have a similar problem. My experience with the piezoelectric crystal type led me to discard this approach and try the beam type of accelerometer.

2) strain gaged beam - The strain gaged cantilever beam has been used extensively in the past. It has the advantage of being easy to damp and is very predictable. It has the disadvantages of low electrical output, and has to have a low natural frequency. As a consequence of the low electrical output, it has a poor signal to noise ratio. The low natural frequency is not a problem in this application. However, the low sensitivity of a strain gage bridge, requires high levels of amplification (i.e. 2000), and this leads to noise problems. Since the sensitivity of the strain gage bridge is directly related to the natural frequency of the beam, the designer is caught between two conflicting requirements, if the natural frequency is to be greater than a few hundred Hertz. This application requires a natural or resonant frequency of around 2 kHz, which automatically means that the sensitivity will be low. Even though an extensive effort was made to solve the noise problem, including an imbedded amplifier chip in the accelerometer carrier near the beam, I realized that low level accelerations (i.e. < 10 G's) would be obscured. Consequently another approach to the beam type of accelerometer was taken.

3) piezo film coated beam - Piezo film is a thin polyvinylidene flouride plastic film, that develops large voltages when stretched or compressed. The thickest film, which was 4.2 mils (100 micrometers), was used. The film is called Kynar, and is produced by Pennwalt Corporation (phone 215-337-6710). Both sides of the film are coated with a metallic coating, which provides an electrical connection. While electrical contact can be made with a silver loaded epoxy, I found that mechanical contact was satisfactory and somewhat easier to do. While it is not as sensitive as piezo crystal materials, it is far more sensitive than a strain gage. For this application an op-amp amplifier with a gain of 50 resulted in an output of several volts at an acceleration of 25 G's. The amplifier is embedded in a small cavity (.35x.75x.2 inches) in the body of the

accelerometer carrier. The signal to noise ratio was very high even at low acceleration levels, because the noise level was only a few millivolts. The output of the piezo film must be fed to the amplifier through a field effect transistor for proper impedance matching. Piezo film has one undesirable characteristic, and that is the electrical-mechanical coupling (i.e. voltage output for a given deformation) changes with frequency. The electrical coupling changes by a factor of two between 50 Hz and about 150 Hz. This frequency range was avoided in this application. Above 150 Hz the coupling is constant.

Accelerometer Design

The accelerometer is deliberately designed to exclude vibration above about 2 kHz. To accomplish this filtering, the beam was designed to have an undamped resonant first mode frequency of 3 kHz without the film. The dimensions of the beam made of steel shim stock are 0.015 inches thick by 0.2 inches wide and 0.40 inches in length. When the film is bonded to both surfaces of the beam and 0.55 of critical damping added, the resonant frequency was lowered to 2 kHz. A photograph of the complete accelerometer mounted on the muzzle of the rifle was previously shown in Figure 4-21. A cross-section drawing of the accelerometer beam is shown in Figure A-1.

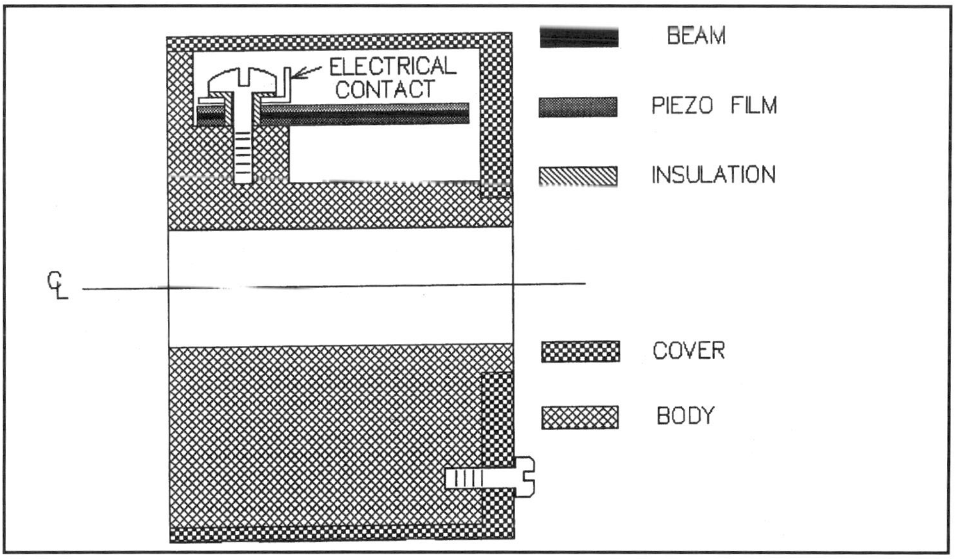

Figure A-1 - Cross-section drawing of the muzzle accelerometer.

The film is bonded to the free end of the beam with epoxy. The electrical contact shown in the figure, plus a connection to the body of the accelerometer provide the circuit connections. The cavity in the top of the accelerometer, that contains the beam, is about 0.35 inches wide, and is filled with silicon oil through a threaded hole in the cover that is not shown. The cover is sealed with Locktight. The silicon oil has a viscosity of 25000 centistokes and is made by Dow Corning. The oil has a consistency about like that of honey, and has to be inserted with a small tube. All of the air must be removed by allowing the device to sit for several hours. The oil provides a damping factor of between 0.5 and 0.6. Any damping fluid that is used must have a very high electrical resistance to prevent charge leakage from the piezo film. The accelerometer is only useful in making dynamic measurements, because the charge on the film will decay in a static environment.

The calculated amplitude ratio and phase characteristics for the accelerometer beam are shown in Figure A-2. Note that the amplitude ratio is flat to within 10% out to about 2 kHz where it starts to roll off, so that the higher frequencies will be attenuated. While the amplitude characteristics are just what are desired, a 40 degree phase lag results at the frequency of most interest (1.25 kHz), and that is not a desirable feature. Fortunately, this can be corrected by designing a band pass filter that will have a similar sized phase lead at 1.25 kHz. Also, the bandpass filter will further attenuate the higher frequencies, which is the main purpose of the filter.

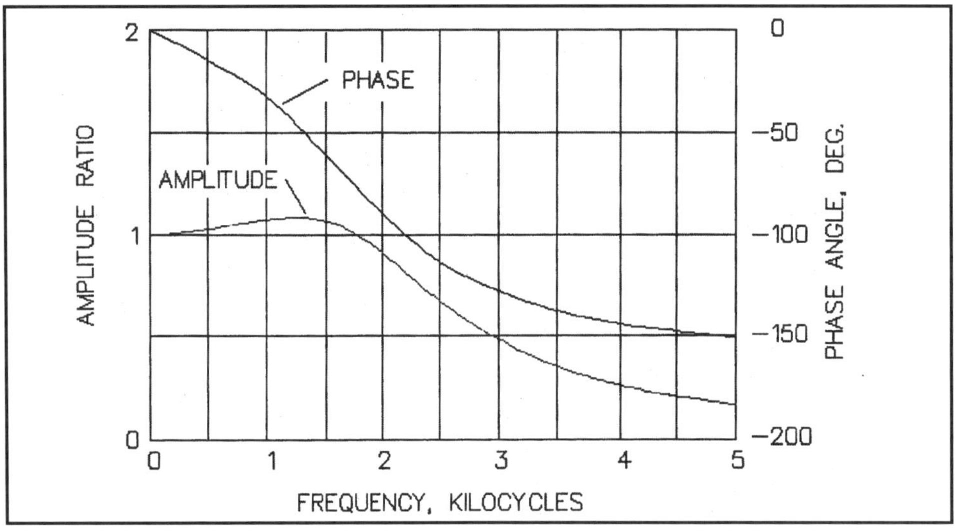

Figure A-2 - Graph showing the frequency response characteristics of the accelerometer.

Bandpass Filter

A bandpass filter was used instead of a low-pass filter, because of its phase lead characteristic at frequencies below the center frequency. The filter is designed to have a phase lead as near 40 degrees at 1.25 kHz as possible, which means that the center frequency has to be somewhat higher than the first mode frequency. The amplitude ratio and phase characteristics are shown in Figure A-3 for the filter that was used. It has a center frequency of 1.4 kHz and a gain of two. The filter was designed to have a Q of two and a bandwidth of 700 Hz. The phase lead is about 35 degrees at the first mode frequency (1.25 kHz), which comes close to compensating for the 40 degree phase lag induced by the accelerometer beam. These calculated response characteristics were confirmed by experimental bench tests. The filter was designed using the equations provided by Berlin in Reference 30. The circuit for the bandpass filter is shown in Figure A-4. Now, the response characteristics shown in Figure A-3 are for steady state conditions, and the input to the filter is not steady state in this application. The only thing to do is to use the differential equation that expresses the behavior of the filter and subject this equation to an artificial, but representative input from the barrel vibration code. In this way a calibration factor can be obtained as the ratio of

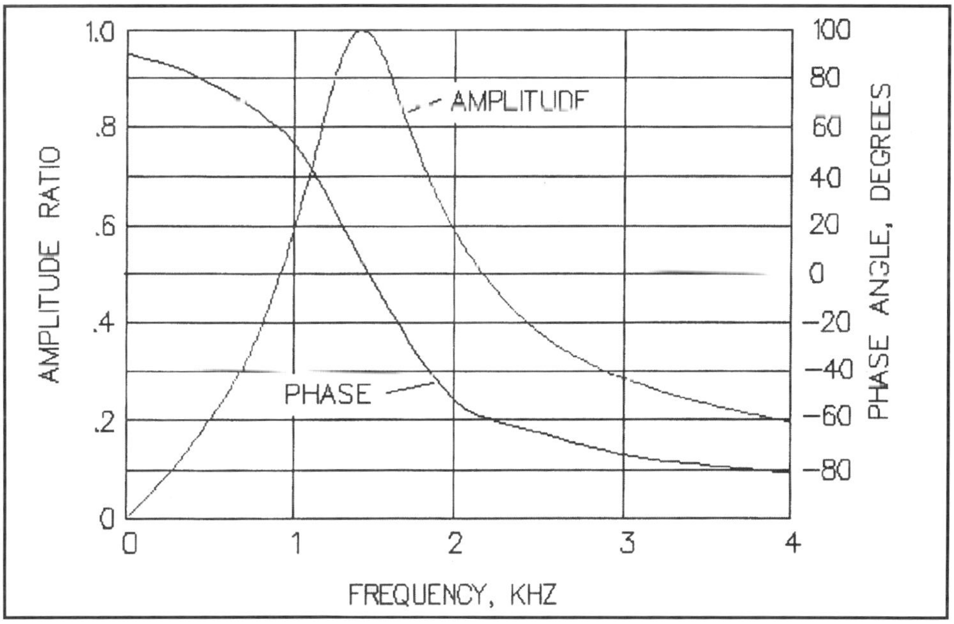

Figure A-3 - Graph showing the frequency characteristics of the bandpass filter used to filter out unwanted signals.

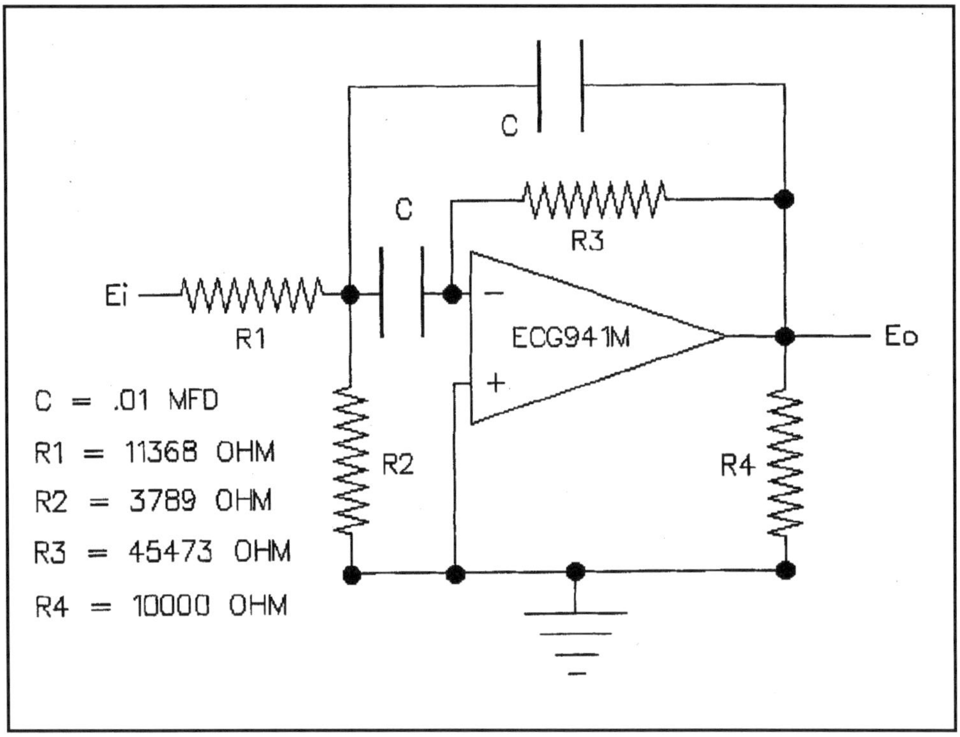

Figure A-4 - Circuit of the bandpass filter.

the input to the output voltage for the typical first mode oscillation of the barrel. The calibration factor turned out to be 0.52 compared to the steady state value of 0.82 at a frequency of 1.25 kHz. The filter doesn't have time to reach an equilibrium steady state condition. Since the second order differential equation used for the transient response calculations seems to be difficult to find in the literature, it is shown below for the benefit of professional investigators.

$$(d^2/dt^2)Eo = -2/(R3*C)*(d/dt)Eo - (R1+R2)/(R1*R2*R3*C^2)*Eo - 1/(R1*C)*Ei$$

where Eo is the output voltage, Ei is the input voltage, and the asterisk (*) indicates multiplication.

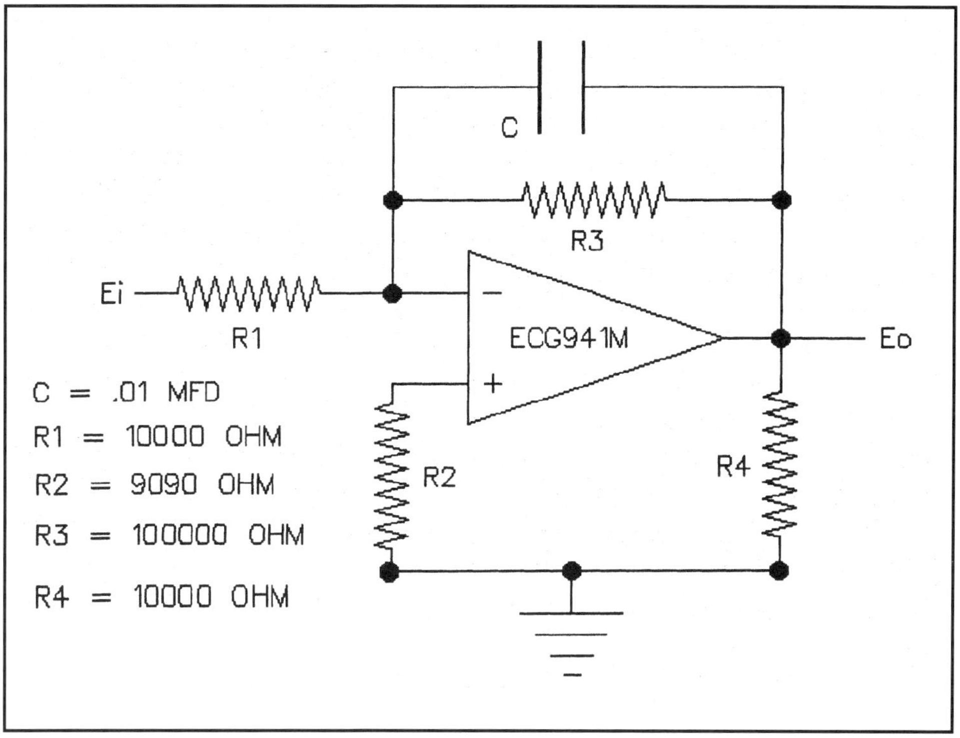

Figure A-5 - Circuit diagram of the integrator used to obtain the muzzle vertical velocity and vertical deflection.

The circuit used for the integrators, which provide lateral muzzle velocity and displacement from the acceleration output of the bandpass filter, is shown in Figure A-5. The output of the bandpass filter is fed to the input of the first integrator, which has an output proportional to velocity, and the output of the velocity integrator is fed to the input of the second integrator, which has an output proportional to the deflection of the muzzle. The gain of the circuit is 10000, which results in a calibration of 0.427 inches/sec/volt for velocity, and 0.043 mil/volt for deflection. Note that screw adjustable potentiometers are used for all resistors, which facilitates exact adjustment. All three variables, acceleration, velocity, and deflection can be compared with the computed variables obtained from the barrel vibration computer code. In addition, integration has a low-pass filtering effect, which makes the experimental velocity and deflection easier to compare and analyze. Also, lateral velocity and deflection are the two parameters that cause bullet dispersion.

Calibration

Calibration of the device begins with vibrating the accelerometer, without the bandpass filter, on the end of a 3/4 inch diameter rod mounted in a lathe. The cross feed of the lathe is used to deflect the end of the rod a known amount, and the carriage feed is used to release the rod so that it vibrates freely. The length of the rod can be varied to change the frequency, however the frequency was generally in the range of 30 to 50 Hz for these tests. The acceleration at the end of the rod can be calculated from

$$A = d*(f/(2*\pi))^2/G$$

where f is the frequency in Hz, G is the acceleration of gravity (32.16 ft/sec^2), d is the lateral deflection in feet, and A is the lateral acceleration in G's. The deflection used was enough to produce 20 to 30 G's. The output voltage is recorded on the oscilloscope for analysis. The results indicated that the sensitivity of the accelerometer is 0.087 volts/G at low frequencies, and is twice that, or 0.174 volts/G at frequencies above about 100 Hz. Recall that the manufacturer provides data showing that the piezo film is twice as sensitive at high frequencies. Of course, it would be better to calibrate the accelerometer at a higher frequency (say 1 kHz), however this requires a professional piece of equipment which was not available, and not sensible for me to try to construct on an amateur basis. The acceleration data from the vibrating rod was also used to check the performance of the barrel vibration code and the agreement was excellent. The accelerometer data on the actual barrel (held in the lathe) was used to obtain a viscous damping coefficient for use in the barrel vibration computer code, by determining the number of cycles required to damp to half amplitude. The damping factor was 0.03, which is typical of mechanical vibration systems.

The total accuracy of the acceleration instrumentation is probably no better than 10%. There are several reasons for this relatively low accuracy.

1) The calibration had to be performed at a much lower frequency than the frequency of interest, because of equipment limitations. Consequently, the calibration depends on the frequency dependence data supplied by the manufacturer.

2) The reading accuracy on the oscilloscope record is probably no better than a few percent.

3) Cross axis sensitivity is the biggest contributer to inaccuracy in the calibration, even though it is undoubtedly much reduced by allowing the accelerometer to slide in the axial direction during the time of measurement. Since the friction force in the axial direction is roughly 1.5 pounds and the accelerometer weighs about 0.16 pounds the axial acceleration acting on the accelerometer should be roughly 10 G's. The cross axis sensitivity was measured and found to be between 5% and 7%. Therefore, the error should be roughly 0.5 to 0.7 G's, which is a relatively small percentage of the peak accelerations (30 G's) that were observed. Since there isn't any way to evaluate this error directly, there isn't any way to be absolutely sure of the error in the actual case.

4) The transient response of the bandpass filter is another source of uncertainty. The transient response of the filter depends on the characteristics of the input, and while the actual input was simulated there is still some uncertainty. The only other thing that might be done is to differentiate the experimental output of the filter (i.e. inverse transform), using the differential equation for the filter, to try to obtain the actual transient response of the filter. This is a difficult undertaking, and was not done.

5) Temperature and other miscellaneous effects were not evaluated.

In spite of the fact that the instrumentation accuracy is not as good as the author would like, it is probably accurate enough to serve the purposes for which it was intended. In fact, 20% accuracy would have been good enough to determine how the muzzle vibrates, and to determine whether or not the corrections made to the rifle were effective in reducing vibration. I believe that the accelerometer instrumentation has been successful in these respects.

B

APPENDIX-B BARREL VIBRATION COMPUTER CODE

This appendix is necessarily written at a higher technical level than the rest of the book, and requires an engineering background for thorough understanding. In spite of this necessary complication, the average reader may wish to scan it for a better understanding of how the barrel vibration code works, and to gain a better comprehension of the complexity and depth of the work.

Several approaches were taken in developing the computer code to simulate barrel vibration. The coupled multiple body approach proved to be the most practical for use on a small personal computer. The barrel is divided into 25 elements or bodies, which is schematically shown in Figure 4-27. The taper in the barrel is much more gradual than that indicated by the computer graph, because of the limited number of CRT pixels. Each element is treated as a constant diameter cylinder. It was found that at least 25 elements per wavelength of the highest mode of oscillation were required. In this case the third mode, which is the one of interest, represents approximately one wave length over the length of the barrel (see Figure 4-28). Attempts to run this code for larger numbers of elements, in order to obtain modes higher than the third, proved fruitless and were abandoned because the computing times were excessive. The author's computer is a Gateway 2000 with an Intel 486 processor running at 66 megacycles. Computing time for a typical case under these conditions is about 6 minutes. The computing time is roughly proportional to the cube of the number of elements, because the time step has to be reduced as the number of elements is increased. Consequently, a main frame computer is required for calculations involving substantially greater numbers of elements. The code simulates vibration in the lateral direction, which is perpendicular to the bore axis, resulting in a single degree of freedom code. Calculations were primarily done in the vertical plane. The single degree of freedom approach should be valid for this application where the deflections are small and rotation of the elements is very small. In applications involving large deflections, it may be necessary to include the effects of rotation. In applications to thick beams, it may be necessary to include the effects of shear deformation, however the barrel is a long slender beam, and the shear effects are not included. The differential equations that describe the motion of each element are

$$A[i,j]*M[j]*d^2Y[j]/dt^2 + Y[i] + C[i,j]*dY[i]/dt = A[i,j]*F[j]$$

where A[i,j] is the influence coefficient for the i'th element with a driving force F[j] at the j'th element, M[j] is the mass of the j'th element, Y[i] is the deflection of the i'th element, C[i,j] is the damping coefficient, dY[i]/dt is the velocity of the i'th element, and $d^2Y[i,j]/dt^2$ is the acceleration acting on the i'th element with a force applied on the j'th element. The equation must be solved for the acceleration, which provides

$$d^2Y[i]/dt^2 = (-Y[i]-\Sigma(\text{inertial})+A[i,j]*F[j]\ )/(A[i,i]*M[i]) \\ -C[i]*dY[i]/dt$$

where

$$\Sigma(\text{inertial}) = \Sigma\{A[i,j]*M[j]*(d^2Y[i,j]/dt^2)\} \\ -A[i,i]*M[i]*(d^2Y[i]/dt^2)$$

The sum of the inertial terms is the sum of the effective forces acting on the i'th element as a result of inertial forces acting on all the other elements. In order to clarify this, we write the equation for the first element, where a single driving force is located at element 1.

$$
\begin{aligned}
d^2Y[1]/dt^2 = \{&-Y[1] + F[1]/M[1] - A[1,2]*M[2]*(d^2Y[2]/dt^2)\\
&-A[1,3]*M[3]*(d^2Y[3]/dt^2) - A[1,4]*M[4]*(d^2Y[4]/dt^2)\\
&-A[1,5]*M[5]*(d^2Y[5]/dt^2) - A[1,6]*M[6]*(d^2Y[6]/dt^2)\\
&-A[1,7]*M[7]*(d^2Y[7]/dt^2) - A[1,8]*M[8]*(d^2Y[8]/dt^2)\\
&-A[1,9]*M[9]*(d^2Y[9]/dt^2) - A[1,10]*M[10]*(d^2Y[10]/dt^2)\\
&-A[1,11]*M[11]*(d^2Y[11]/dt^2) - A[1,12]*M[12]*(d^2Y[12]/dt^2)\\
&-A[1,13]*M[13]*(d^2Y[13]/dt^2) - A[1,14]*M[14]*(d^2Y[14]/dt^2)\\
&-A[1,15]*M[15]*(d^2Y[15]/dt^2) - A[1,16]*M[16]*(d^2Y[16]/dt^2)\\
&-A[1,17]*M[17]*(d^2Y[17]/dt^2) - A[1,18]*M[18]*(d^2Y[18]/dt^2)\\
&-A[1,19]*M[19]*(d^2Y[19]/dt^2) - A[1,20]*M[20]*(d^2Y[20]/dt^2)\\
&-A[1,21]*M[21]*(d^2Y[21]/dt^2) - A[1,22]*M[22]*(d^2Y[22]/dt^2)\\
&-A[1,23]*M[23]*(d^2Y[23]/dt^2) - A[1,24]*M[24]*(d^2Y[24]/dt^2)\\
&-A[1,25]*M[25]*(d^2Y[25]/dt^2)\}/(A[1,1]*M[1]\\
&-C[1]*(dY[1]/dt)/M[1]
\end{aligned}
$$

So, we have to solve 25 of these equations simultaneously for the acceleration on each of the 25 elements. This is done on a computer by simply putting the following equation in a loop that sums the terms over j for each i.

$$
\begin{aligned}
\Sigma(\text{inertial}) = \Sigma\{&A[i,j]*M[j]*d^2Y[j]\}\\
&A[i,i]*M[i]*(d^2Y[i]/dt^2),\ j=1 \text{ to } 25 \text{ for all i's}
\end{aligned}
$$

Then one subtracts $A[i,i]*M[i]*d^2Y[i]/dt^2$ from the sum of the acceleration forces and plugs the result into

$$d^2Y[i]/dt^2=\{-Y[i]-\Sigma(\text{inertial})+A[i,j]*F[j]\}/(A[i,i]*M[i]) - C(i,j)*dY/dt/M[i]$$

This equation is solved for each i element. It should be noted that I found that the solution was greatly stabilized by multiplying the summation of acceleration term by a factor slightly less than one (i.e. 0.99 to 0.995). The reason for this is unknown, but may be due to the limited 32 bit word length of the small computer. A fourth order Runge-Kutta integrator is used to integrate the accelerations to obtain velocities and then the velocities are integrated to obtain displacement. The new values for acceleration, velocity, and displacement are substituted back into the equations and the equations are solved again for new values of the variables (i.e. state variables). This process is repeated until the accelerations stop changing significantly, at which time the solution has converged. At this point the solution continues to the next time step. The number of iterations was arbitrarily limited to 50 after some experience was gained in operating the code. The time interval for each step was chosen to be one microsecond. This is adequate for the frequency of the stiffest element (i.e. element 1) at the fixed end of the barrel. Matrix solution schemes were not tried, because the solution used was satisfactory. Successive Over Relaxation (SOR) was used to speed up the convergence. A SOR convergence factor of 1.2 seems to be optimum. The constant convergence criteria used was usually in the region of 0.06 G, however this depended on the individual problem. A better convergence criteria might be to have a variable criteria applied that depends on the stiffness of the individual elements. The boundary conditions are simply zero for all state variables.

In the general case one can have applied forces at all elements, however in this specific case the recoil and other receiver moments are applied at element 1. In solving for the gravity droop a constant one G acceleration is applied at all elements, and the solution is allowed to continue until the barrel stops moving. The deflection at each element due to gravity is saved in an initial condition file for starting the code for the usual vibration solution. The gravity droop has little effect on dispersion. It can have an effect on the point of impact for different elevation angles of the rifle. As the rifle is elevated from the horizontal, the gravity droop reduces, causing the bullet to strike higher on the target. Other special conditions were investigated, such as the centrifugal force caused by the bullet traveling along the curved path due to gravity droop, the effect of centrifugal force developed by a spinning unbalanced bullet, barrel stiffening due to pressurization, and others. None of these effects appeared significant, and are not shown in the equations in order to avoid excessive complication.

The influence coefficients (i.e. A[i,j]), provide the coupling terms between the elements. They are the static deflection caused at the i'th element by a unit (1 pound) force applied at the j'th element, and in effect are the reciprocal of the usual spring constant. They are solved for by simply solving the static deflection equations, which are

for i>j
dY[i,j] = dx*dx^2/(6*EI[i]*(3*(j-1)+2)
dY[i,j]/dx = dY[i-1,j]/dx + dx^2/(2*EI[i]*(2*(j-1)+1))
Y[i,j] = Y[i-1,j] + dx*dY[i-1,j]/dx +dY[i,j]
A[i,j] = Y[i,j]/12

for i≥j
dY[i,j]/dx = dY[i-1,j]
Y[i,j] = Y[i-1,j] + dx*dY[i,j]/dx
A[i,j] = Y[i,j]/12

where x is the distance along the barrel from the fixed end, E is Young's Modulus (30 million psi), I is the moment of inertia of the cross-section, dx is the length of the element (1 inch), Y is the vertical deflection, and A is the needed influence coefficient. Since Y will be in inches with the usual units for I and E, Y is divided by 12 to obtain A in feet/pound. The reader should note that a force, including inertial forces, acting at any element influences the motion of all other elements. This is what couples the bodies together.

The stiffness of the first element, which is the forward receiver ring, was determined experimentally by using strain gages on the ring. It was found that the stiffness (i.e. E*I) was about 70% of the calculated value. This is not surprising, because the ring is a complicated structure making calculations difficult. Also, no cantilever beam ever has complete end fixity, and this one is no exception. In calculating the influence coefficients, the mass is assumed equally distributed along the length of the element.

In addition to the main part of the code, the moments resulting from bolt and recoil force are computed from a table of chamber pressures obtained from experimental measurements. The recoil moment was obtained from a two body spring mass model, where the first body consists of the barrel, action, and scope, which is connected by a spring to the second body consisting of the stock. The stock spring constant was determined to be in the region of 100,000 pounds/inch with analytical methods. A value of 96,000 pounds/

inch proved to give good results in the code. This spring constant represents the compression of the wood stock behind the recoil lug as the recoil force is applied to the stock. It has the effect of delaying the onset of the applied moment to the receiver. These applied moments are then input as a force acting at the end of element 1 with a moment arm of one inch. The response moment can be determined from the deflection at the first element using the beam equation. The response moment is, of course, the moment that is measured by the strain gages, and is less than the applied moment. The longitudinal position of the bullet was also input in table form from the data in Chapter 2. There is also graphic coding required to plot the barrel and its deflection. Since graphic coding is sensitive to the particular computer involved, it is not presented.

The code was successfully checked by comparing with muzzle acceleration obtained from bench vibration tests on both a 3/4 inch constant diameter rod and the actual barrel. These tests were described in Appendix A.

Several hundred computational runs were made with this code over a period of two years for the purpose of studying the effect of various inputs on barrel vibration. It was invaluable in researching the causes of barrel vibration and finding out just how the barrel moves, and how much dispersion is caused by the movement. The barrel vibration code has not been documented, and is not user friendly, although it could be. Consequently, it is not available to the public at the present time.

C

APPENDIX-C
BULLET BALANCE DEVICE DESIGN

A dynamic balance device is preferable to a static balance device, be cause it not only measures the CG lateral displacement (i.e. offset), but can be used to measure the principal axis tilt with respect to the geometrical axis. The principal axis is the axis about which a projectile will spin in free flight. The dynamic balance device is usually more dependable and sensitive than a static balance device. However, the static balance device is easier to make and requires less equipment to operate.

Static Balance Device

Design

The static balance device shown in Figure 9 4 (Chapter 9) is nothing more than a torsional pendulum. A cross- section drawing of the device is shown in Figure C-1. The outside dimensions of the device are 2.625 inches long, 0.75 inches high, and 0.50 inches wide. The steel wire (0.01 dia.) extends about 2 feet on either end from the carriage, and is placed under tension. The wire was made from a steel guitar string. The carriage is made of aluminum and the sides are highly polished to reflect a light beam. The axial hole is first drilled through the aluminum block, and polished so that the bullet makes a close slip fit in the hole. The end holes are enlarged to accept the wire end fittings at this point. The end fittings are made to have a close fit with the

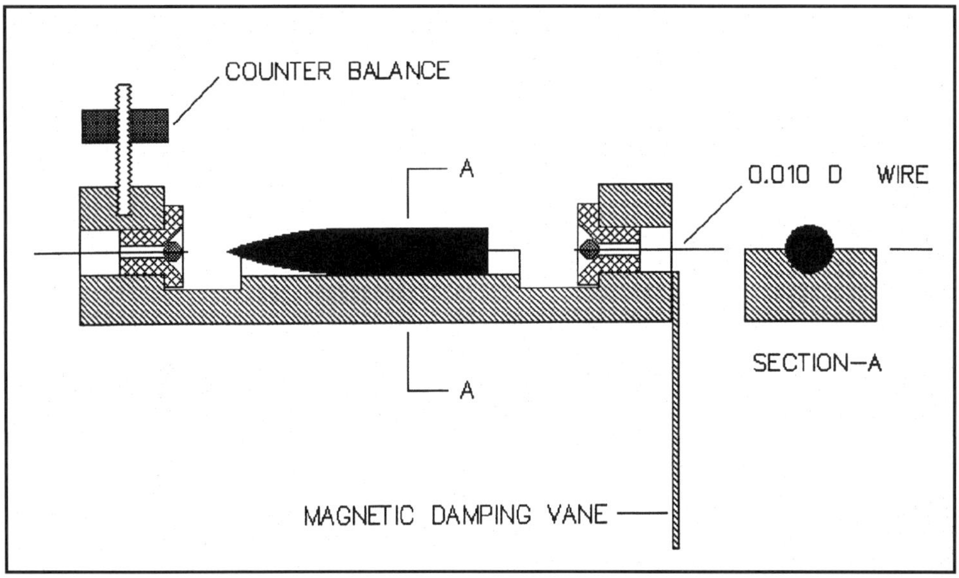

Figure C-1 - Drawing of the static balance device.

holes. Then the top of the carriage is milled away, and the edges deburred. The silver solder beads on the ends of the wires are formed by holding the wire in a vertical position, and melting a small drop of solder onto the end of the wire. Gravity will cause the drop to form a perfect tear drop shape. The 0.032 thick aluminum damping vane is epoxy bonded to the carriage. The vane travels between two small magnets placed on both sides of the vane near the end of the vane. The brass balance weight is adjusted so the CG of the device is near the centerline of the bullet. The carriage will rotate first to the right and then to the left as an unbalanced bullet is rotated in the carriage. This rotation is detected by shining a light beam onto the highly reflective side of the carriage, which when reflected provides a spot of light on a screen placed in a perpendicular direction some 20 feet away. The up and down excursion of the light spot is measured to determine the amount of CG offset. The light beam can be formed by shining a small, high intensity light into the eyepiece of a rifle scope.

Operation

I mount the device in a lathe, because it is a stable, convenient way of holding it. The wire supports must be stretched tight, so that the carriage does not sag. The higher the CG of the carriage is, the greater the sensitivity. However, if one raises the CG of the device too high, it will become unstable and dump the bullet. To check the balance of a bullet, the bullet is rotated about

45° at a time, and the position of the light spot is noted. The maximum up and down excursion of the spot is recorded, and can be converted to CG offset in inches by multiplying the result times a calibration constant. The calibration constant is determined by drilling a radial hole halfway through the bullet at the longitudinal CG position. The CG offset of the calibration bullet is easily calculated, and should be large enough (i.e. 0.001 inch) to overwhelm any existing CG asymmetry before the hole was drilled . A more accurate approach is to make a brass cylinder that has the same weight as the bullet to be tested, and drill a hole in it to unbalance it. That way the possibility of picking a badly balanced bullet for a calibration bullet is avoided. I used a well balanced bullet obtained from the dynamic balancer described below. The calibrated sensitivity of this particular device, for an 18 foot optical path length, is 0.02 mils per inch of total spot deflection. That is, the total spot deflection is the deflection between highest and lowest positions of the light spot. Figure C-2 shows the results of balancing the same 100 bullets that are balanced with the dynamic balancer described below, and it can be seen that while the general result is the same as that obtained from the dynamic balancer, the static balancer results are not as smooth.

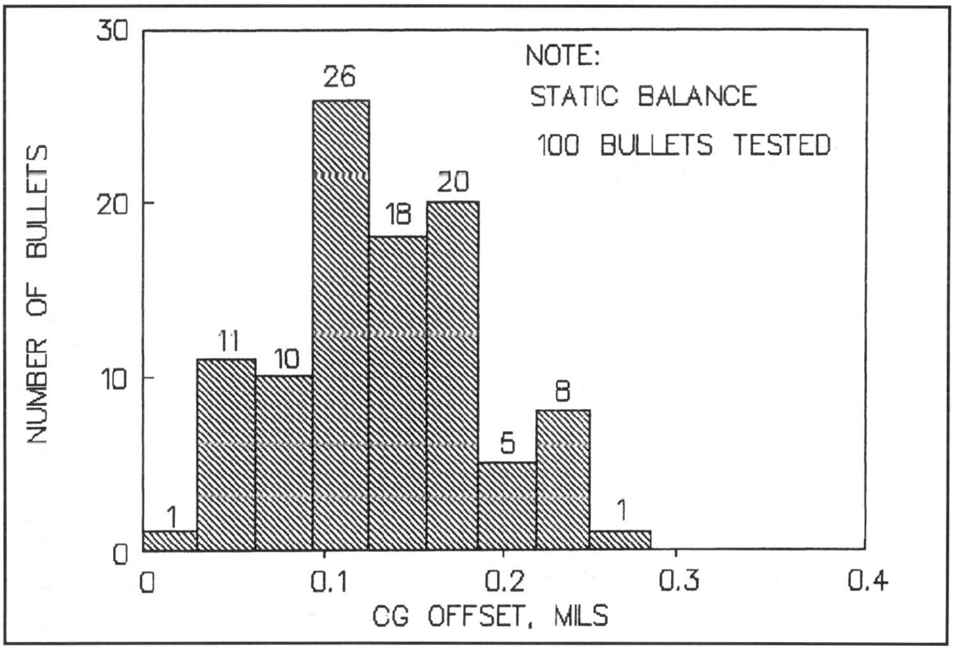

Figure C-2 - Measured center of gravity offset of 100 bullets obtained from the static balance device.

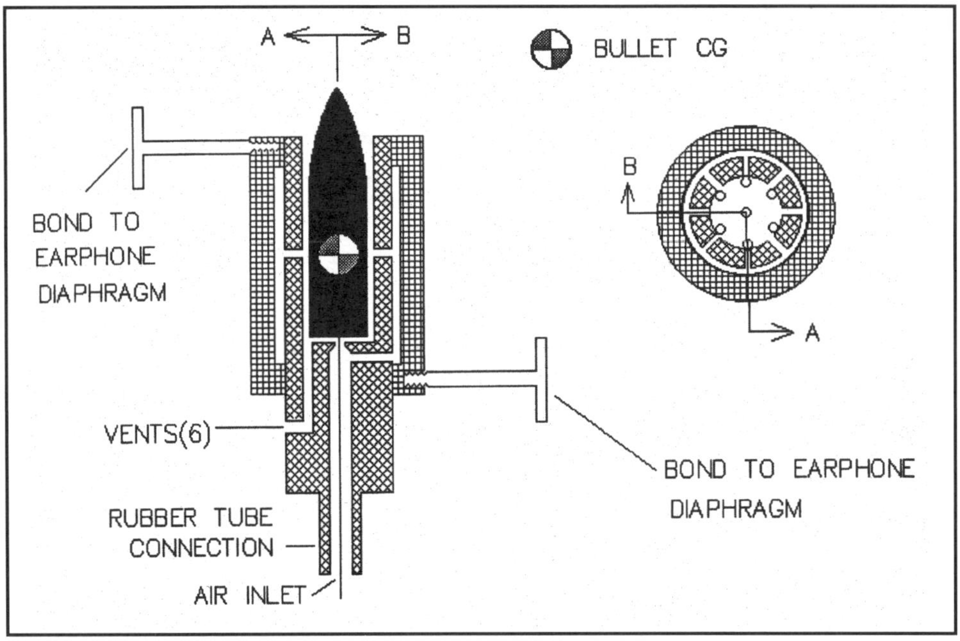

Figure C-3 - Drawing of the dynamic balance device showing the air passageways that allow the bullet to be suspended and spun on an air cushion.

Dynamic Balance Device

Design

The bullet dynamic balance device design is based on the use of air bearings to constrain the bullet as it spins at a high speed (≈150 cps). A photograph of the balance device is shown in Figure 9-6 (Chapter 9), and a cross-section sketch is shown in Figure C-3. The sketch is not drawn exactly to scale. The device is made of polystyrene (plexiglass), a transparent plastic, to facilitate drilling all the air passages. Machining must be done at a slow rate with water coolant to prevent melting of the plastic. There are two air bearings involved. One air bearing supports the base of the bullet, and the other provides lateral support along the sides of the bullet and provides the spin action. Air enters the bottom of the device through a small orifice and spreads laterally until it exits through the six equally spaced vent holes near the sides of the cavity. This air cushion, between the base of the cavity and the base of the bullet, lifts the bullet a few mils preventing contact between the base of the bullet and the device. The high pressure air in the circumferential chamber squirts through the eight equally spaced radial holes (0.016 inch dia.) and flows out through the six base vents (.052 inch dia.) and the top of the cavity. The air in the circumferential chamber is supplied through two diametrically placed holes (0.052 inch dia.) into the air supply passage (0.125 inch dia.). One of these supply holes is shown in Section B. The other passage into the circumferential chamber cannot appear in Section A. The eight radial holes are drilled parallel to the diameter, but are offset laterally about 20 mils to provide the spin action. This offset is not shown in the sketch. The cylindrical cavity for the bullet is tapered slightly, measuring 0.279 inches at the bottom and 0.282 inches at the top for a bullet diameter of 0.277. This taper greatly improves vertical stability of the bullet. The taper is obtained by hand finishing with fine emery paper and finally polishing with a fine polishing compound on a wood dowel. I have found that tooth paste makes a good polishing compound. The device should be designed so that the CG of the bullet will be half way in the vertical direction between the two mounting posts that connect to the earphone diaphragms. The posts are bonded to the diaphragms with epoxy cement. The two plastic cylinders are bonded with plastic cement. The outside diameter of the external cylinder is 0.65 inches, and the height of the outside cylinder is 0.60 inches. The reader is cautioned that the dynamic balance device requires careful machine work on good equipment to be successful. The earphones are of the old magnetic type, and do not require any electrical power.

Operation

A dual sweep oscilloscope is required if one wishes to measure principal axis asymmetry, however a single sweep oscilloscope is sufficient to measure CG offset. The oscilloscope is required to check the spin frequency, because the voltage output is proportional to the square of the spin frequency. Usually, the frequency of reasonably well balanced bullets will be very consistent, and all one has to do is check to be sure. Badly balanced bullets, such as those used in calibration, may not spin as fast as the regular bullets, and the voltage will have to be increased by the square of the spin rate ratio. A calibration bullet, or bullets, are made by finding a well balanced bullet and drilling a small radial hole to the center line at the longitudinal CG position. The edge of the hole must be carefully deburred. In the data shown for the 90 grain Sierra 270 bullet (Figure C-4), a hole diameter of 0.065 inches provides a 1 mil offset, which is as large an offset that one should try. In practice the oscilloscope is used to check the spin frequency and a digital multimeter set on AC is used to measure the voltage. The calibration of this device yielded 0.508 volts (RMS) per mil of CG offset, which is a convenient level. The

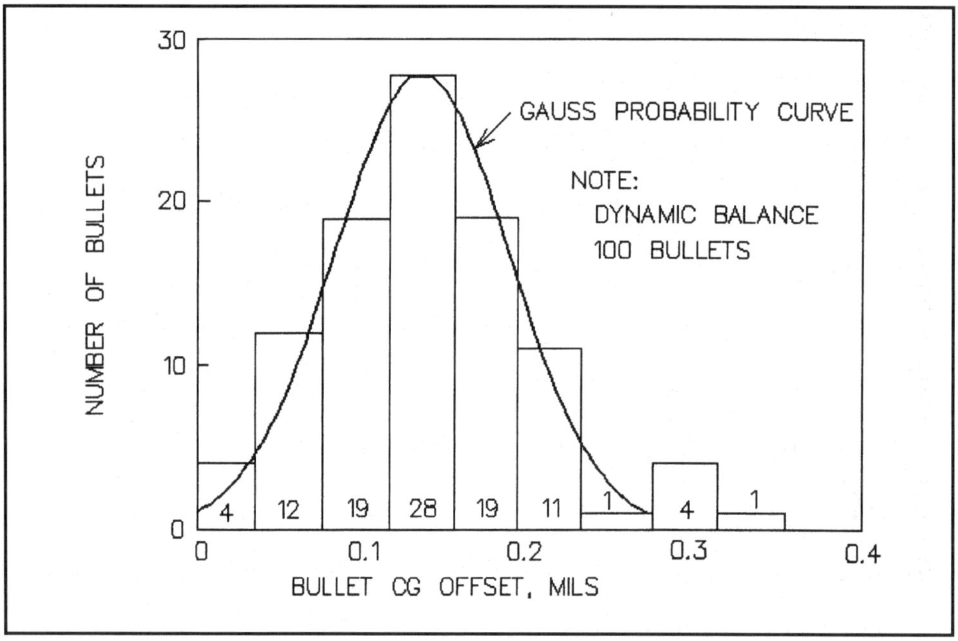

Figure C-4 - CG offset data for 100 bullets obtained from the dynamic balance device compared to a standard Gausian distribution.

operator has to be careful to allow plenty of time for the bullet to spin up to a steady state level (143 cps). This may require as much as two minutes, so it is a slow process. In practice the spin rate will usually over speed at first, then oscillate and gradually come to a steady rate. If the spin rate is not steady the inlet air pressure may be a little too high. In this case the bullet is rising and falling in the chamber. With experience, one can usually tell when the spin rate has reached equilibrium by simply listening to the high pitched hum that the device generates. I use a parallel-jaw machinist clamp on the rubber inlet tube to adjust the air flow that comes from the compressor, which is regulated at 40 psi. A small air compressor is sufficient to provide enough high pressure air. Obviously, the bullets and the chamber must be kept clean, because the clearance between the sides of the bullets and the chamber are small. In all probability the voltage output of the two earphones will not be equal, because the sensitivity of earphones vary. Simply adjust one of the traces so that both signals are identical. The voltage output of the two earphones are exactly 180 degrees out of phase. If one sums these two voltages, the difference voltage will be proportional to the principal axis asymmetry. This means, of course, that the oscilloscope must have at least one summing amplifier if principal axis tilt is to be measured. After working with this device for a long time I have never seen a significant principal axis asymmetry. I believe this is inherent in bullet production methods. The principal axis, which is coincident with the CG, is just shifted laterally without being rotated with respect to the center line axis, or axis of geometrical symmetry. This probably results from unequal jacket thickness from side to side. Consequently, the reader may not want to bother with trying to measure the principal axis asymmetry.

Resonance between the bullet spin rate and the lateral oscillation frequency of the device is one thing that should be guarded against, because resonance effects may cause an error. However, the damping of the earphones is very large, which inhibits resonance effects. The natural frequency of the device can be determined by simply tapping the side of the device and watching the output on the scope. I found in this case that the frequencies were a little to close, so I wrapped a 1/2 inch wide strip of 1/16 inch thick sheet lead around the cylinder between the two support posts. This lowers the frequency of the device sufficiently to avoid resonance after spin up has occurred.

D

APPENDIX-D
SIX DEGREE OF FREEDOM (6DOF) COMPUTER CODE

The first Six Degree Of Freedom (6DOF) computer code was developed by NASA around 1960. These early 6DOF codes required long computing times on large main frame computers. These days they can be run efficiently on personal computers with clock speeds of 20 MHz or faster. The 6DOF code computes the trajectory of a projectile exactly, provided that the aerodynamic coefficients and the mass characteristics are accurately known. However, they should be operated by experienced flight dynamicists for reliable results. The 6DOF code allows the projectile to rotate and translate in three mutually perpendicular planes. These are all the possible components of motion, consequently the motion is rigorously defined by the equations. Most of the trajectory data that is published for rifle bullets comes from a very simple point mass computer code, which completely ignores rotation of the bullet. This automatically eliminates the effects of bullet asymmetries, spin, and transient motion, and consequently, the point mass trajectory codes are useless for accuracy studies. The point mass approach is adequate for determining mid-range trajectory height and velocity at various ranges, which is helpful to the shooter. The point mass approach is also much faster and vastly easier to use as a result of its simplicity. Since we are going to investigate the effect of several things on external ballistic accuracy (see Chapter 10), the 6DOF code is required.

Equations Of Motion

The 6DOF code works by solving six equations of motion which describe the way the bullet behaves in the six degrees of freedom. This particular code can be simplified by making the assumption that the trajectory angles are small. The trajectory angles are small for a bullet trajectory over short ranges (i.e. 300 yards or less). Also, it is assumed that the aerodynamic coefficients are constant with Mach number, which is also true for the typical supersonic rifle bullet over reasonable ranges. Both of these simplifications speed up the computation.

The three equations that describe the rotational motion about the roll (x), pitch (y), and yaw (z) axes are

dp/dt = Mx/Ixx
dq/dt = (-p*r*Ixx + My)/Iyy
dr/dt = (p*q*Iyy + Mz)/Izz

where p,q, and r are the roll, pitch, and yaw angular rates, Mx, My, and Mz are the roll, pitch, and yaw aerodynamic moments, and Ixx, Iyy, and Izz are the moments of inertia of the bullet about the roll (x), pitch (y), and yaw (z) axes. The quantities on the left side of the equations are the angular accelerations about the roll, pitch and yaw axes. When integrated these accelerations provide the angular rates of the bullet (p, q, and r), and when integrated a second time provide the three attitude angles of the bullet.

The three lateral translational equations that act along the three roll (x), pitch(y), and yaw(z) axes are

du/dt = r*v - q*w - g*sinΘ + Fx/m
dv/dt = r*tanΘ*w - r*u + Fy/m
dw/dt = q*u - r*tanΘ*v + g*cosΘ - Fz/m

where du/dt, dv/dt, and dw/dt are the translational accelerations, and Fx, Fy, and Fz are the applied forces acting along the x, y, z axes. The m is mass of the bullet. If the bullet is launched in a near horizontal direction, so that the pitch angle Θ is small and the range is not too large (300 yards), these equations simplify to

$$du/dt = r*v - q*w - g*\Theta + Fx/m$$
$$dv/dt = r*\Theta*w - r*u + Fy/m$$
$$dw/dt = q*u - r*\Theta*v + g - Fz/m$$

These three accelerations can be integrated to provide the velocities along the three axes. These velocities can be used in the following equations to determine the flight path of the bullet relative to an earth fixed coordinate system.

$$dXe/dt = u*\cos\Theta*\cos\Gamma - v*\sin\Gamma + w*\sin\Theta*\cos\Gamma$$
$$dYe/dt = u*\cos\Theta*\sin\Gamma + v*\cos\Gamma + w*\sin\Theta*\sin\Gamma$$
$$dZe/dt = -u*\sin\Theta + w*\cos\Theta$$

The translational equations can be simplyfied for small angles and are

$$dXe/dt = u - v*\Gamma + w*\Theta$$
$$dYe/dt = u*\Gamma + v + w*\Theta*\Gamma$$
$$dZe/dt = -u*\Theta + w$$

These simplifications are only desirable when using the older personal computers (PCs), such as the Z80 and the 80186 without a coprocessor. The three Euler angles in roll, pitch and yaw (ø, Θ, and Γ) are obtained by integrating the following angular rate equations.

$$d\o/dt = -r*\tan\Theta$$
$$d\Theta/dt = q$$
$$d\Gamma/dt = r*\sec\Theta$$

These angular rate equations can be simplified to

$$d\o/dt = -r*\Theta$$
$$d\Theta/dt = q$$
$$d\Gamma/dt = r$$

A fourth order Runge-Kutta integrator is used for integrating all equations. The aerodynamic angle of attack and angle of sideslip are required for determining the aerodynamic forces and moments in the six equations. These are

$$\alpha = \arctan(w/u)$$
$$\beta = \arctan(v/u)$$

The aerodynamic force and moments must either be calculated or obtained from experimental data. In general I have used experimental aerodynamic and inertial data in calculating the external ballistics for this book. A minimum of two force coefficients and four damping coefficients are required in the case of a rotationally symmetric projectile. I think I have gone about as far as possible in explaining to the average reader how these complex calculations can be made. The 6DOF computer code will tell you in great detail how a bullet will fly, if you have the correct aerodynamic and inertial data, and the correct launch conditions. This has been proven over and over by flight tests.

E

APPENDIX-E TUNNEL RANGE CONSTRUCTION

After fighting with wind drift problems for years and finding it difficult to separate wind effects from other accuracy problems, I decided to build a Tunnel Range to eliminate wind effects. With the approval of the Zia Rifle and Pistol Club of Albuquerque we built it on the Club Range. The club purchased enough used four foot inside diameter concrete pipe to make a 100 yard range for five hundred dollars. My son and son-in-law laid in railroad ties which were surveyed with a rifle scope on a tripod to within 1/2 inch. The ties were placed about six feet from the ends of each section of pipe. Most of the pipe sections were 34 feet long and weighed about 12 tons each. I rented two 30 ton cranes and one flatbed truck for $1700 - one crane for loading at the salvage yard and one for unloading at the site. A cross was made of 1x2 lumber that would fit in the far end of the pipe and was used as a crosshair for alignment. This worked very well. The final alignment required one inch shims on only two ties to bring the centers of the pipe sections within one inch of alignment throughout its length. Tapered chocks, made from 6x6 lumber, were driven between the railroad ties and the pipe to prevent any motion of the pipe. The chocks were nailed in place with large nails. Figure E-1 shows the range before being covered with about 500 cubic yards of dirt to maintain thermal equilibrium. An 11x15 foot concrete slab was poured at the entrance end of the tunnel and a building constructed on the slab. The forward end of the building is made of concrete block and acts as a retaining wall for the dirt covering. A photo of the finished range is

Figure E-1 - Tunnel Range showing exposed pipe before covering with earth.

shown in Figure E-2 and a view of the interior of the building is shown in Figure E-3. The outside of the 2x4 structure is covered with 3/8 inch plywood. The roof of the building is covered with corrugated translucent fiberglass. In the summer 4x8 foot sheets of thin white polyethylene foam are nailed to the bottom of the ceiling rafters to provide shade. A 3.5 inch thick shooting bench was made of reinforced concrete poured in place on the concrete filled concrete block pedestals. However, we found the concrete block pedestals too unstable and later replaced them with one foot diameter solid cast concrete pedestals. The pedestals were cast in commercial cardboard

Figure E-2 - Finished range. Louvered inlet for fan is to the right of the door.

Figure E-3 - Photo of the interior of the building showing concrete bench and entrance of the 100 yard tunnel.

forms and proved to be perfectly rigid. The building was wired for electric outlets and lights. A target light was installed. This made it possible to conduct the muzzle blast studies at night. A loading table is shown on the left and an adjustable stool is by the bench. There is a gate at the muzzle end that serves to keep out trash when the tunnel is not in use and is locked in the open position to prevent people from wandering in between the target butt and the tunnel muzzle. All of the muzzle end, with the exception of an opening for the muzzle gate, is surrounded by a berm. The target is seven feet from the muzzle end of the tunnel, which allows room for a chronograph. The bench was placed so that the rifle muzzle would be about six feet from the tunnel entrance. The Oehler 35P chronograph gates with lights are placed inside the mouth of the tunnel on short stands made of 2X4 lumber. There is a louvered opening to the right of the building door that conceals a 30" diameter exhaust fan. This facility has proved to be a very nice tool, however mirage has proven to be a problem.

Before building the tunnel range I had talked to several people with some experience with them, but no one seemed to say much about mirage problems, which proved to be nearly as bad as the wind. The first time three of us tried firing in it we all got groups with nearly pure vertical dispersion.

When we watched through the scope we could see the reticule move up or down over a total distance of as much as 0.6 inches. One of the shooters (Bill Minneman), that I had talked to, mentioned that he had installed an exhaust fan to reduce the mirage effect, so I tried installing a fan that pulled the air from the muzzle end toward the entrance at a rate of about 7 feet per second. The fan reduced the slow mirage drift by about a factor of three

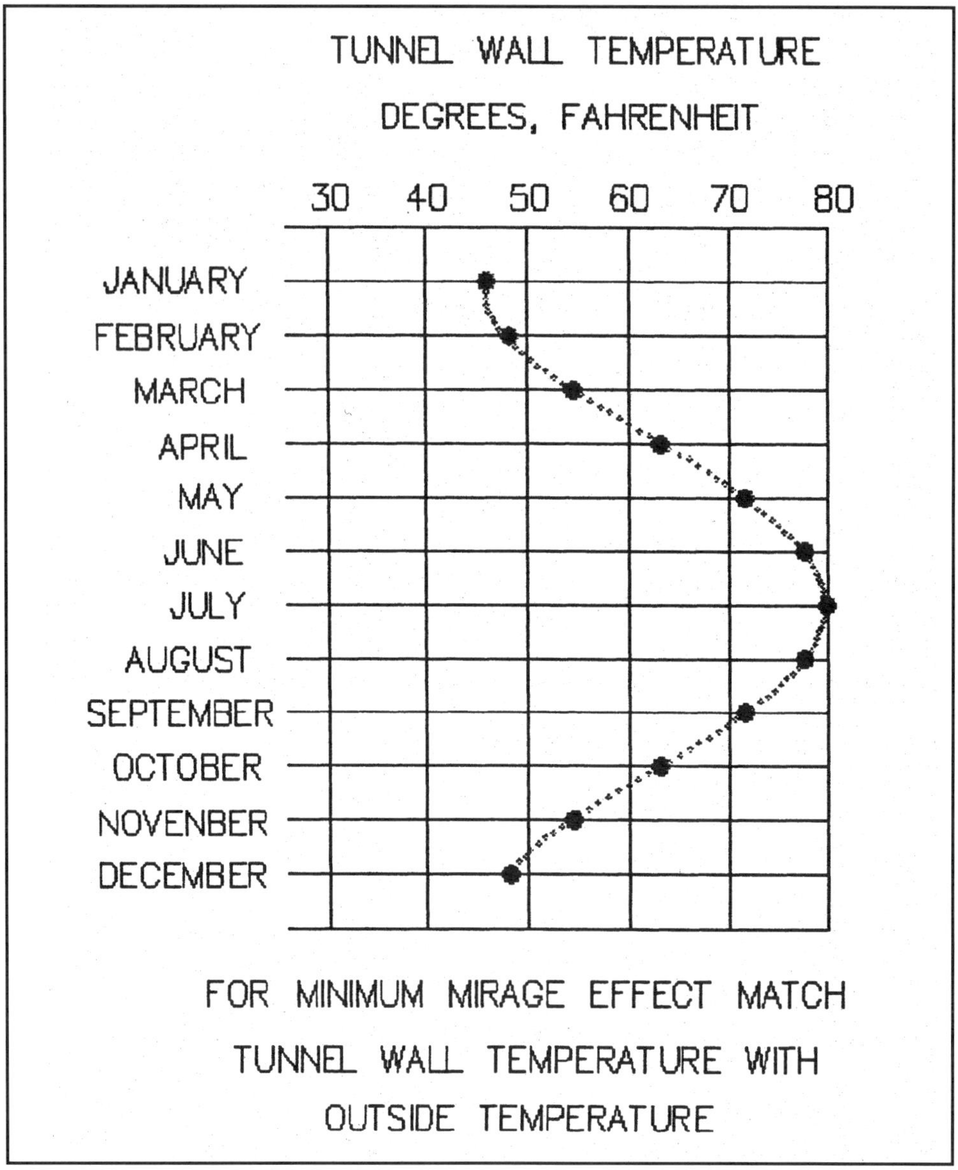

Figure E-4 - Graph showing variation of Tunnel wall temperature for the months of the year.

(i.e., 0.2 inch), which is still too much when you are trying to test a rifle that shoots in the ones and twos (i.e., 0.1 to 0.2 inch groups). I found that it worked much better if the air was forced down the tunnel, which meant sealing the building. Meanwhile I had been measuring the tunnel wall temperature (Figure E-4) and found that if one matched the outside free air temperature with the tunnel wall temperature the mirage disappeared. A thermometer was hung on the shady side of the building to indicate free air temperature and a copy of Figure E-4 was posted in the building. Since we are in a desert climate, the tunnel wall temperature change from winter (49°F) to summer (80°F) is much larger than you would expect in some climates. The tunnel temperature essentially follows the average daily temperature. Before turning on the fan the top and bottom wall temperatures usually differ by 3°F to 4°F. After the fan has run for 30 minutes the surface temperatures at the top and bottom of the tunnel seem to reach equilibrium. On the Pacific coast the seasonal temperature doesn't vary much and one should have less trouble with mirage. The next improvement was to build a reference scope mount (Figure F-1) that holds a scope with the reticule initially fixed on the aim point as a reference. The rifle scope is then aimed at the same point that the reference scope indicates should be the true aim point without mirage. My experience indicates that with the fan on and the temperatures matched within 5°F the correction due to mirage is less than 20 mils on the target while firing a single group. This seems to have solved the mirage problem. The reference scope mount holds the reference scope at the same level as the rifle scope and about two inches to the left, so that the shooter can move rapidly back and forth between scopes for comparison. I also tried projecting a laser spot on the target that was about 0.25 inches in diameter at the target and shooting at the laser spot. However, this was not too successful because scintillation caused the spot to twinkle like a star and made aiming difficult. With a rail gun that holds its zero between shots a reference scope really isn't necessary. However, I usually use the reference scope because it not only checks mirage conditions but tells you if the rail gun has moved between shots.

The Tunnel Range may not work for transonic velocity (1000 to 1500 fps) projectiles because the normal shock waves will be reflected back from the tunnel walls to the bullet. This can cause instability of the bullet with large dispersion. We know that it doesn't work for low or medium large caliber bullets such as muzzle loader or pistol bullets, because we have seen oblong bullet holes in the target. However, it may work for 22 RF if the trajectory is in the center of the tunnel and low speed match ammunition is used. The 22 may be small enough compared to the tunnel inside diameter that the shock waves could be too weak to effect the bullet. A tunnel range is just not worthwhile for anything but accurate guns.

If I built another one of these ranges about the only thing that I would change would be to place the fan in the middle of the building so that it lined up with the tunnel center line. Since the fan is off center, it tends to generate a swirling motion of the air about a vertical axis in the building, which may cause some dispersion. I might also consider placing a vertical baffle between the fan and the rear of the bench, because it gets cold sitting in the draft when the outside temperature drops below 55°F. One might also consider building a small building at the muzzle end of the tunnel for safety reasons. We had to build this range so that it pointed toward the south to conform to the Zia Range layout. Unfortunately, the wind often is out of the south or southwest which interferes with the flow induced by the fan. It would be best to orient the tunnel to point in a direction parallel to the prevailing wind. We only have an annual rainfall of 8 inches, so drainage is not a problem.

I have found the Tunnel Range to be essential in doing rifle accuracy diagnostic work because it eliminates the worst variable - wind effects. As far as mirage effects are concerned they are present on an open range but shooters are often unaware of them. The tunnel range exaggerates mirage effects but by using the fan and matching free air and tunnel wall temperatures the mirage is effectively eliminated. However, one should use the reference scope to be sure of the mirage conditions.

F

APPENDIX-F RAIL GUN

I have made a mount to hold a sporter stock, which was shown earlier in Chapter 2 (Figure 2-2). It was not as accurate as a regular rail gun but served its purpose of holding the instrumented sporter. I decided I needed a rail gun for testing in the Tunnel Range and the gun shown in Figures F-1 and F-2 was the result. Also shown in Figure F-1 is a reference scope and mount for the reduction of mirage problems. Both scopes are 36X Bauch and Lomb.

Figure F-1 - Photograph showing the left side of the rail gun and the mirage scope on a separate mount. Note counterweights fastened to the barrel block.

Figure F-2 - Photograph of the right side of the rail gun.

I started out with a used gun to save time and money. On the original gun the front "T" shaped bearing mount was on a separate tripod and the rear adjustment mount was on a separate plate. I mounted both of these on a 3/4" thick aluminum base plate, which made the gun a lot more rigid. The Bakelite and Delrin bearings were replaced with Teflon, which I consider much superior. The measured static friction coefficient with the Teflon bearings was 0.025, which is very low. The foot screws are 1/2" diameter and are secured with 1/4" cap head set screws. Carbide inserts from 30 caliber armor piercing bullets are silver soldered into the end of the foot screws. These carbide inserts are hard and sharp enough to penetrate a concrete bench top with a gentle tap from a small hammer, so they stay put. The procedure in setting the base is to set the front of the base on two short pieces of 1x2 wood and move the base until the sight is pointed at the aim point. Then the two front foot screws are turned until the front of the base is level and above the two wood blocks. Then the lock nuts on the two foot screws are tightened and tapped into the bench with a small hammer. The next step is to tighten the set screws that lock the foot screws. The procedure is repeated on the rear foot screw. The rails are cleaned and lubricated with Pledge, a furniture polish or Friction Block.

Figure F-3 - Photograph of the rear of the rail gun showing details of the windage and elevation adjusting mechanism.

The original windage and elevation adjustment on the rear of the gun was unsatisfactory and was modified to prevent movement between shots. A photo of the final adjustment system is shown in Figure F-3. The 3/4 inch steel adjustment plate is hinged with a close fitting 3/8 inch diameter steel rod toward the front of the plate. The capstan wheel underneath the rear of the plate provides elevation adjustment and the two screws on either side provide windage adjustment. There are two vertical hex head bolts at the front of the adjustment plate that stabilize the plate by providing a preload. The procedure is to aim the rifle about two inches above the intended aim point then tighten the two hex bolts finger tight and raise the rear of the adjustment plate with the capstan wheel until the elevation is correct. The horizontal adjustment is done with the two horizontal screws and both are tightened. As far as I can tell this gun does not move between shots and with a 36 power scope you can detect very small movement of the reticule. If there appears to be any movement I check for mirage effects by comparing the mirage scope position to the rifle scope. In this way you know whether the gun has moved or you are observing a mirage effect.

I tried several different barrel block designs and finally settled on aluminum V-blocks as being the best.These different methods included steel V-blocks, aluminum blocks with circular cuts to match the barrel, and epoxy bedding. The bottom block is bedded in Devcon F epoxy and is held by six 3/8 inch bolts. The position of the blocks is adjusted with shims until the carriage can be moved all the way with no more than 1 mil variation on a dial gage attached to the base. Six 3/8 inch bolts clamp the top and bottom blocks together and are torqued to 50 inch pounds. The V-block surfaces are coated with a solution of rosin in acetone to prevent any possibility of movement.

There are 3/4x4x7 inch steel blocks bolted to both sides of the top barrel block (Figure F-2). These are counterbalance weights that counteract the recoil moment caused by the recoil force acting on the blocks. When the gun is fired the rearward recoil force is opposed by the forward inertial force acting on the carriage. These two forces cause a couple (moment) tending to rotate the barrel blocks and the barrel in a muzzle upward direction. The forward inertial forces acting on the two counterweights acts to compensate the recoil moment reducing the moment acting on the barrel. This was determined by measuring the moment acting on the barrel just forward of the barrel blocks with strain gages. Figure F-4 and F-5 show the measured

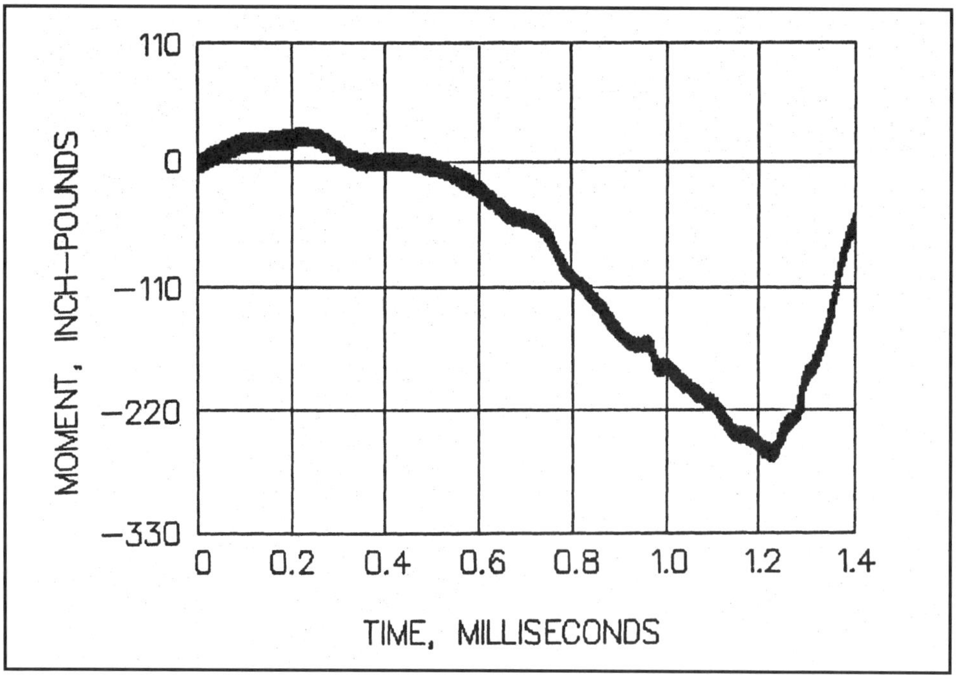

Figure F-4 - Computer scan of oscilloscope data on rail gun barrel moment without counterbalance weights.

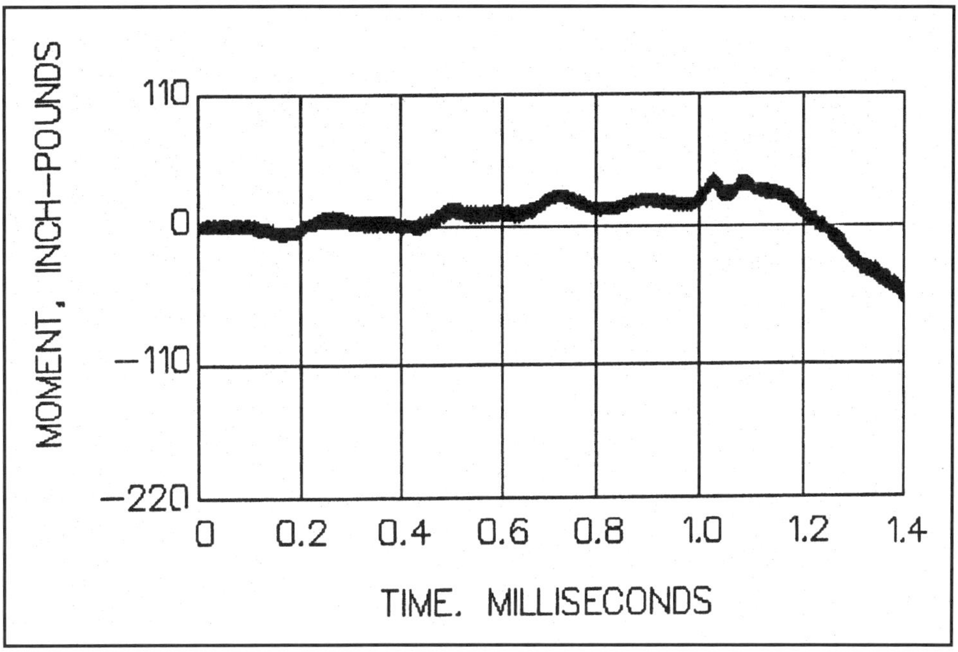

Figure F-5 - Computer scan of oscilloscope data showing rail gun barrel moment with counterweights. When compared with Figure F-4 it can be seen that the weights greatly reduced the moment acting on the barrel, thereby reducing vibration.

moments with and without the counterbalance weights. You can see that the moment is reduced to near zero with the counterweights. This greatly reduced the vertical stringing that was present in the groups without the weights. See Chapter 4 for a discussion of high frequency vibration problems observed on this gun.

There is a 1.875" OD aluminum tube that covers the barrel ahead of the barrel blocks. This tube, which was suggested by Frank Tirrell, helps to maintain a constant circumferential temperature of the barrel which minimizes thermal distortion and shifting of point of impact due to differential cooling.

About the only thing that I can think of to improve this gun would be to lower the carriage so that it is closer to the base plate. It has proved to be a very handy tool for evaluating ammunition problems. The carriage and the base each weigh approximately 45 pounds.

If I were to build one of these things from scratch I would consider using flexures like those used on the Recoil Isolator in Chapter 4 instead of bearings. The carriage only has to recoil about 0.010 inches before the bullet exits the muzzle which should be easy to accommodate with flexures. The recoil could be absorbed by an adjustable hydraulic damper. However, the amount of horizontal windage adjustment might be somewhat limited. I would also try to design the carriage so that the CG of the carriage plus barrel and action would end up on the bore line. This would prevent some vibration problems.

G

APPENDIX-G SHADOWGRAPH TESTING

Shadowgraphs have been used for years in diagnostic testing in ballistic ranges and wind tunnels. A high intensity, short duration beam of light, usually from an arc light source, is passed through the flow region of interest and a shadow of the flow is cast on a sheet of film which is exposed. The image shows significant density gradients such as turbulence and shock waves. It allows an investigator to really see a physical picture of the flow region.

I should warn the reader that the high voltage and energy involved in the arc light source is exceedingly dangerous and very likely will kill you if you accidentally contact the high voltage. It is much more dangerous than the high voltage in a television set because of the high energy involved. So, I advise against duplication of this equipment unless the reader is experienced in working with high voltage equipment.

Figure G-1 shows the equipment setup with the rail gun in the Tunnel Range. The white box in the foreground contains the high voltage supply and the trigger electronics. In the background on the right side you can see the arc head, which is enclosed in a Nylon box. The Nylon knob on the right hand side of the arc head adjusts electrode spacing. The microphone is visible near the muzzle that triggers the electronics. On the left side of the gun a black screen can be seen. The 12"x18" lithograph sheet film is clamped to the black screen. The round black device in front of the box containing the

Figure G-1 - Photograph of the shadowgraph equipment set up in the Tunnel Range with the rail gun. The round object in the foreground is a variac that adjusts the line voltage feeding the high voltage power supply. The white cabinet on the right contains the electronic circuitry. Behind the white cabinet is the arc light. The black board on the left holds the lithographic film.

electronics is a variac, which adjusts the line voltage to the high voltage power supply. The high voltage is indicated by the microammeter on the right front of the electronics box.

The spacing between the arc head and the muzzle is about 12" and the distance between the muzzle and the film screen varies from 4" to 10" depending on whether the large sheets of lithographic film (12"x18" Fuji GA-100) or Type 57 Polaroid 4x5 film is used. The 4x5 camera using the Polaroid film is mounted above and behind the arc head. The Polaroid film is exposed at f-4.5 and is useful in checking the result, but not very good for reproduction. You can quite often see the image with the eye well enough without photography to tell whether are not the timing is correct. The lithographic film is developed in Kodak Tmax developer for 7-10 minutes. The photographs shown in the book were made by photographing the 12"x18" film negatives on a light table with Tmax 100 film in a 4x5 camera and then reversing with a Kodak Tmax 100 reversal developing kit. The resulting negatives were printed.

Figure G-2 shows a front view of the setup, showing the trigger microphone near the muzzle for late time shots. Figure G-3 shows the hole in the side of the barrel used for early time shots. The hole is 3/16"D at the surface and the last 0.05" is 1/16"D. The hole is 5.5 inches from the muzzle.

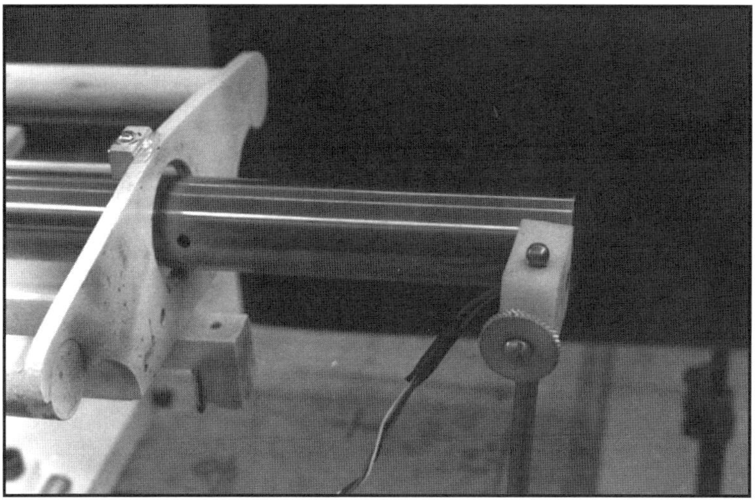

Figure G-2 - View of the side of the barrel showing the microphone at the muzzle and the hole in the side of the barrel for early time triggering. The microphone is placed near the hole for early time triggering and the time adjusted by moving the microphone away from the barrel for more time delay.

Figure G-3 – Front view of the shadowgraph setup, showing the arc light source on the left and the film mounting board on the right. The microphone is placed near the muzzle for late time shots.

You can see where the powder smoke has discolored the white surface of the gun carriage. Put some kind of baffle between the hole in the barrel and the operator to avoid being hit in the face by powder fragments.

The total time delay in this system averages about 0.1 msec. Most of the time delay is in the reed relays in the flip flop circuit. To operate, one pushes the reset button and a red LED will come on indicating the unit is ready to fire. When the microphone receives a sound pulse it is amplified and triggers a flip flop. The flip flop controls a power transistor which controls the primary circuit in an auto ignition coil. This collapsing field in the primary of the ignition coil induces a high voltage in the secondary, which is connected to the trigger arc electrode in the arc head. When the trigger arc fires this causes the main arc to fire and discharge the capacitor bank in the arc head. When the unit fires a yellow LED comes on indicating the unit is on standby. If the unit is left in the ready condition too long the high current (3 amps) may cause the power transistor and the ignition coil to overheat. The unit can be fired manually by pushing the test button when the red LED is on (ready position). The microphone is a small capacitor type purchased from Radio Shack. The arc head design is an adaptation of a design used at Sandia National Laboratory which was furnished by Dr. Ken Cole. Commercial versions of this device are available from EG&G Electro-Optics, Salem, MA (508-745-3200).

The discharge time of this device is about 0.5 µsec, but since most of the light energy comes out in about 2/3 that time the effective light pulse is about 0.32 µsec. The bullet moves about 0.012 inches in that time. So the motion is effectively stopped. The high voltage supply is 10 kv that charges six 0.02 mfd capacitors in parallel. **Again, do not trust the insulation or the current bleed circuit to make this thing safe. It is deadly!!!!**

GLOSSARY & ABBREVIATIONS

acceleration	The response of a body to an applied force.
accuracy	The ability of rifle to fire bullets into a target near the aim point.
afterbody	The rear portion of a bullet shape that starts just behind the nose section. May include a boat tail.
ballistic coefficient	The ratio of the bullet mass to a function of the aerodynamic drag force.
balloting	Erratic side to side or angular motion of a projectile in a gun bore.
base	Refers to the rearmost surface of a bullet.
blowby	Powder gasses that travel around the bullet before the bullet enters the bore.
bullet seating depth	The depth that the bullet is seated into the case neck. Also refers to the depth that the bullet is seated into the throat.

burning rate	The rate that gun powder burns at a given pressure in inches per second.
caliber	The maximum diameter of a bullet in inches or millimeters.
cannelure	A circumferential groove impressed into the afterbody of a bullet.
coning motion	The motion a bullet makes with its nose traveling in a circle while the CG remains fixed on the flight path.
center of gravity	That point in a body where the mass can effectively be assumed to be concentrated. Also center of mass.
chamber	The cavity in a rifle barrel that contains a cartridge up to the start of the throat or leade.
chronograph	An electronic instrument that measures the velocity of a projectile.
compression stress	The force that tries to compress a piece of material divided by the area perpendicular to the direction of the force.
diametrical clearance	Difference in diameter between two concentric circles.
drag coefficient	The coefficient formed by dividing the aerodynamic drag force by the dynamic pressure and bullet cross section area.
dynamic pressure	The pressure caused by the motion of a gas. Equal to one half the gas density times the square of the velocity.
expansion shock	The shock wave cased by the lateral expansion of the muzzle blast.

external ballistics	The study of the flight of a projectile after it leaves the influence of the barrel. Also flight dynamics.
extreme spread	The difference between the lowest and highest muzzle velocities of a group of bullets and the dimension between the centers of the widest bullet holes in a group.
feed-back	Applies to an electronic circuit where some of the output current or voltage is fed back to the input of the circuit increasing the amplification.
free run	The distance a bullet must travel before it contacts the throat in a chamber.
friction force coefficient	The ratio of the force required to slide two pieces of material to the force holding the two pieces together.
grain	Unit of weight. There are 7000 grains in one pound.
group size	The distance between the centers of bullet holes that have the largest spread in a group. Also precision.
headspace	The space between the face of the rifle bolt and the head of the cartridge case.
Heavy Varmint (HV) rifle	A bench rest target rifle weighing up to 13.5 pounds.
heel	The corner of the bullet at the base.
internal ballistics	The science of predicting the behavior of the bullet inside the barrel and the forces and stresses on the barrel.
kilocycle	One thousand cycles per second.

Light Varmint (LV) rifle	A bench rest target rifle weighing up to 10.5 pounds.
mass	The weight of an object divided by the gravitational acceleration.
microsecond	One millionth of a second.
millisecond	One thousandth of a second.
Mach disk	A flat surface shock wave normal to the flow velocity where the flow is Mach one.
Mach number	The ratio of velocity to the speed of sound in a gas. Named for Professor Mach in the 1930's.
muzzle ventilation	The practice of drilling holes in a barrel near the muzzle to relieve the muzzle blast pressure.
moment	The product of a force times the distance between the force and point of application
normal shock wave	A planar shock wave that forms perpendicular to the direction of the gas flow. The flow behind the shock wave is Mach one.
oscilloscope	An electronic instrument that displays the variation of a signal voltage on a cathode ray tube similar to a television tube.
parallax	An optical problem in a telescopic sight where the image appears to move when the eye is moved off the optical axis of the scope.
precursor shock wave	A shock wave formed at the muzzle by the compressed gas traveling ahead of the bullet.
pressure ring	A small oversize ring on the heel of some bullets produced during manufacture.
radial clearance	Difference in Radius between two concentric circles.

rail gun	A rail gun has a barrel and action clamped to a carriage which slides on rails that ride on bearings. They usually are heavy (100 pounds) and are used for test purposes, although they are used in bench rest unlimited class competition.
run out - (RO)	The measurement taken on the surface of a cylinder with a dial gage that is rotated about a longitudinal axis not necessarily on its center line.
secant ogive	A bullet nose shape generated by a segment of a circular arc that is not tangent at its intersection with the afterbody.
shadowgraph	The practice of shining a high intensity short duration light through a flow field onto a sheet of film to get a photograph of the flow including shock waves.
shock wave	A shock wave represents a sharp discontinuity in pressure, density and temperature that travels through a gas (air). A sound wave is a very weak shock wave that travels at the speed of sound in air (1160 fps). The pressure, density and temperature is higher behind the wave.
slug	Unit of mass - pounds/gravitaional acceleration (G)
strain gage	A thin metal foil that changes electrical resistance when stretched allowing a measurement of strain.
tangent ogive	A bullet nose shape generated by a circular arc section where the arc is tangent to the afterbody.
tension stress	The force that tries to stretch a piece of material divided by the area perpendicular to the direction of the force.

thermistor	An electrical resistor that changes electrical resistance with a change in temperature.
throat	The tapered entrance just ahead of the chamber where the bullet enters the barrel. Also, leade.
transition ballistics	The study of the behavior of a projectile as it leaves the muzzle of a barrel but is still in the influence of the muzzle blast.
twist rate	The distance along a barrel that it takes for the rifling or a bullet to make one revolution.
ultimate strength (stress)	The stress where a piece of metal breaks.
yield strength (stress)	The stress level where a piece of metal starts failing and will no longer return to its original shape when the load is removed.

Abbreviations

cps	cycles per second
fps	feet per second
G or g	acceleration of gravity, 32.16 feet per second^2
kc	kilocycles, thousands of cycles per second
mm	millimeter, 1/1000 of a meter, 1/25.4 inch
mil	one thousandth of an inch
ms	milliseconds, one thousandth of a second
rad	radian, equals 57.3 degrees
sec	second
μsec	microsecond, one millionth of a second
°F	degrees Fahrenheit
°c	degrees centigrade
°R	degrees Rankine - degrees Fahrenheit + 459.6
CG	center of gravity
GS	gyroscopic stability factor

REFERENCES

1. "Absolute Chamber Pressure In Center Fire Rifles", 1965 by Brownell, York, Sinderman, Jacobs, and Robins. University of Michigan, Ann Arbor, Michigan

2. "Theory Of The Interior Ballistics Of Guns", 1950 by Corner. John Wiley and Sons, New York.

3. "Gun Propulsion Technology", Vol. 109 Progress in Astronautics and Aeronautics, American Institute of Aeronautics and Astronautics, Washington, DC 20024

4. "Pressure Measurements In The Transitional Ballistics Region Of A M-16 Rifle", 1975 by Gion, BRL Report No. 1765, USA Ballistic Research Laboratories, Aberdeen Proving Ground, Maryland.

5. "The Effect Of Muzzle Jet Asymmetry On Projectile Motion", 1975 by Schmidt, BRL Report No. 1756, USA Ballistic Research Laboratories, Aberdeen Proving Ground, Maryland.

6. "The Intermediate Ballistic Environment Of The M16 Rifle", 1976 by Zoltani, BRL Report No. 1860, USA Ballistic Research Laboratories, Aberdeen Proving Ground, Maryland.

7. "Investigations Of The Transitional Ballistics In Muzzle Jet Flow Simulators", by Oertel, BRL Report No. 2686, USA Ballistic Research Laboratories, Aberdeen Proving Ground, Maryland.

8. "The Prediction Of Gun Muzzle Blast Properties Utilizing Scaling", by Fansler and Schmidt, BRL Technical Report No. ARBRL-TR-02504, USA Ballistic Research Laboratories, Aberdeen Proving Ground, Maryland.

9. "Interior Ballistics Of Guns", 1979, Vol. 66, Progress in Astronautics and Aeronautics, American Institute of Aeronautics and Astronautics, Washington, DC 20024

10. "Advanced Gunsmithing", by W.F.Vickery, 1940, Kingsport Press, Kingsport, Tennessee.

11. "Gunsmithing", by Roy F. Dunlap, 1950, Small Arms Technical Publishing Company, Georgetown, South Carolina.

12. "Machinery's Handbook", 1975, Industrial Press, 200 Madison Ave., New York, NY 10016.

13. "Comparison Of Computed And Measured Jump Of 120mm Cannon", 1990 by Schmidt et al, Sixth USA Symposium On Gun Dynamics, Taimiment, Pennsylvania.

14. "Launch dynamics Of Fin-Stabilized Projectiles", 1989 by Schmidt et al, AIAA-89-3395, AIAA Atmospheric Flight Mechanics Conference, Boston, Massachusetts.

15. "Investigations On The Dynamics Of Tank Guns", 1990 by Bornstein et al, Sixth USA Symposium On Gun Dynamics, Taimiment, Pennsylvania.

16. "Flexible Projectile Modeling Using The Little Rascal Gun Dynamics Program", 1990 by Erline et al, Sixth USA Symposium On Gun Dynamics, Taimiment, Pennsylvania.

17. "The Flexure Of A Uniformly Pressurized Circular, Cylindrical Shell", by J.D.Wood, ASMR Journal Of Applied Mechanics, Dec. 1958 (p453)

18. "An Introduction To The Design And Behavior Of Bolted Joints", J.L.Bickford, 1981, Marcel Dekker Inc., New York, NY

19. "Mechanical Engineering Design", J.E.Shigley, 3rd Edition, McGraw Hill, p250-252.

20. "Numerical Investigation Of Inviscid Shock Wave Dynamics In An Expansion Tube", Keun-Shik Chang, Jong-Kwan Kim, Shock Waves (1995) 5:33-45.

21. "The Bullets Flight", Dr.F.W.Mann (1856-1916), Copyright 1909, Reprinted 1980 by Wolfe Publishing Co., PO Box 30-30, Prescott, Arizona 86302.

22. "A Detailed Development Of The Tricyclic Theory", H.R.Vaughn, 1968, SC-M-2933, Sandia National Laboratories, Albuquerque, NM

23. "Free Flight Motion Of Symmetric Missiles", C.H.Murphy, Ballistic Research Laboratories Report No. 1216, 1963, US Army Ballistics Research Laboratories, Aberdeen, Maryland.

24. "The Aerodynamic Characteristics Of The 7.62MM Match Bullets", R.L.McCoy, Ballistics Research Laboratory Memorandum Report BRL-MR-3733, 1988, US Army Ballistics Research Laboratories, Aberdeen, Maryland.

25. "A Magnus Theory", H.R.Vaughn and G.E.Reis, 1973 American Institute of Aerodynamics and Astronautics Paper No.73-124, AIAA 11th Aerospace Sciences Meeting, Washington, DC.

26. "Walter Watts' Wind Machine", Walter Watts, The Rifle Magazine, July-August 1969.

27. "Aerodynamic Data For Small Arms Projectiles", W.Braun, Ballistics Research Laboratories Report No.1630, 1973, US Army Ballistics Research Laboratories, Aberdeen, Maryland.

28. "Gun Tubes", US Army Material Command, AMC Pamphlet 706-252.

29. "The Aerodynamic Characteristics Of .50 Ball, API, M8, And APIT, M20 Ammunition", R.L.McCoy, 1990, Ballistics Research Laboratories Report No.3810, US Army Ballistics Research Laboratories, Aberdeen, Maryland.

30. "Design of Op-Amp Circuits", H.W. Berlin, Howard W. Sams and Company, 1984

NOTES